GOLD UNDER BRACKEN

the land of Wales

GOLD UNDER BRACKEN

the land of Wales

Richard Hartnup

y Lolfa

First impression: 2011

Design: Y Lolfa
Cover photograph: Cader Idris from Llanfachreth
Richard Hartnup
Author photograph: Faith Purkis

ISBN: 978 184771 314 8

FSC
Published and printed in Wales
on paper from well maintained forests
by Y Lolfa Cyf., Talybont, Ceredigion SY24 5HE
website www.ylolfa.com
e-mail ylolfa@ylolfa.com
tel 01970 832 304
fax 832 782

Contents

	Acknowledgements	7
	Preface	9
1	Landscapes and Soils	12
2	Estuaries	27
3	Sand Dunes	37
4	Alluvial valleys & terraces	47
5	Lowland pastures and woods	59
6	Ceunentydd – the deep gorges	72
7	Heaths	81
8	Ffridd Lands – The Lower Hills	88
9	Lowland bogs and rhosydd – seepage bogs	94
10	Limestone landscapes	102
11	Moorlands and upland bogs	107
12	Mountains	113
13	Artificial Soils	121
14	What did soils ever do for us?	124
	Glossary	127

Tal-y-llyn/Llyn Mwyngil

Aur dan y rhedyn;
Arian dan yr eithin;
Newyn dan y grug.

Gold under bracken;
silver under gorse;
starvation under heather.

Traditional Welsh proverb

Acknowledgements

Many friends and colleagues have read and commented on my drafts for this book, and I thank them all. Especial thanks go to Andrew Agnew, Jan Martin and Gerald Morgan, who gave much appreciated encouragement as well as practical advice, and to Inga Peck and Michael Hayes, who read the text to each other and came up with several useful suggestions. Arthur Chater also made many useful suggestions and put me right on some matters of fact. Kathryn Alton of the British Soil Science Society and Michelle Clarke of the National Soil Resources Institute are among a number of my soil colleagues who supported this project and looked over parts of the text.

I am very grateful to the editors and staff of Y Lolfa for undertaking to publish and promote a book which is not in the mainstream. It helps to adjust the balance as there are so few popular, adult texts that deal with the vital resource that we call soil.

Finally, I would like to applaud the unstinting support and practical help given by my wife Bethan who insisted that I rewrite over-technical or (can I admit it?) less than exciting passages. Her frankness was always tempered by tact, and a faith in my ability to improve.

Richard Hartnup
March 2011

Carneddau Hengwm, with Llŷn Peninsula in the background

Preface

People and land are the two greatest resources of any country. This book is a celebration of the landscape of Wales and its influence on people and ways of life. Coal miners, quarry workers, farmers and craftsmen and women have formed the backbone of the people and culture for the last two hundred years, and other forms of mining go back to the Bronze Age. The remnants of prehistoric farming communities are to be found in abundance, and in practically all types of landscape. The aim here is to look beyond the merely picturesque and to investigate the basis or framework of the landscape – the combinations of soil, climate, geology and land use that make it what it is. I have heard it said that to analyse the beauty of landscape is to negate its spiritual worth. I profoundly disagree with this idea. In more than 40 years of exploring the Welsh countryside, I have come to love and respond to it in ever deeper and more satisfying ways as I have begun, albeit in a small way, to understand more of what underlies our experiences of beauty and learning.

What is it that moves us to respond to beautiful countryside? Answers to this deceptively simple question go beyond aesthetic considerations into the roots of psychology and culture. Landscape is experienced with all of the senses – the feeling of wind and sun, of grass or fallen leaves beneath the feet; the smell of peat, sea or gorse; and the sounds of birds and insects – all have inspired great literature and art. Combinations of form, or shape, and colour stimulate artistic creativity. But it is the half-concealed meanings and the deeper significances of the land that I find most exciting. For millennia people have used the natural resources of Wales for an astonishing variety of purposes, and the successive layers of their influence remain in the landscape to be interpreted. In the now less intensively used uplands, traces of occupation – living, worshipping and communicating – go back to scarcely imaginable times before the last Ice Age. Cromlechs, upright stones and cairns stand alongside intricate patterns of walls and hedge banks that still function for today's farmers. The hillside terraces of prehistoric cultivators stand out in the low sun. These human features of the landscape are moulded to the natural patterns of hill valley and river with an often satisfying sublety, sensitivity and utility.

The light in western Britain changes from minute to minute and from day to day, and variations of humidity give a unique quality to that light, whether it be the clear-washed brilliance that bathes the land after storm clouds clear, or the wreaths of mist that create the mystery of a Japanese print – even on the most unpromising of days. The rains that soak our fields and forests on such a regular basis are responsible for the startling greenness of the fields, and the lush mysteries of the deep moist gorges with their rich carpets of ferns and mosses which frequently conceal the works of our ancestors. These qualities have inspired many

Winter light at Llanfachreth

artists to create their own interpretations of the land, and how it speaks to us today.

Although it is this light that illuminates the countryside, it is the soil that controls much of the nature of the landscape itself. The nature of the soil determines the combinations of plants that can be present, and it is the vegetation that gives colour and texture to the earth's surface. So here I will concentrate on the relationships between soils and vegetation and attempt to give a picture of the land as a whole. The study of soil has, in some quarters, had the reputation of being dull. I hope that this book will dispel this outrageous idea by linking soil types, and their amazing variety, to a myriad other aspects of the land of Wales. We will look at their vital importance to so many facets of our lives. After all, we depend on the stuff for much of what we do – we build our houses, schools, roads and factories on it, play games, walk in the countryside, enjoy gardens and parks; and, most important of all, we rely on it for the provision of food, fibres, timber and biofuel. It is these last vital uses of soil that have driven our ways of understanding it. Thus far, most of the texts concerning the soils of Wales emphasize agriculture and forestry. Whilst I shall draw heavily on the science that has derived from this approach, I hope to give a more general and approachable account of the land of Wales by looking at soil in relation to ecology, landscape and land use. The book is aimed at the general reader who would like to know how the soils vary across the Welsh landscape, and how they affect our use of the land. It will also supply useful information to those studying geography and biology on A Level and undergraduate courses. I do not intend to give a comprehensive scientific inventory of the soils of Wales. That has been done by the Soil Survey and Land Research Centre (now the National Soil Resources Institute) and those who wish to go into more detail should refer to *Soils and their Use in Wales* by C C Rudeforth *et al.* I would strongly recommend the soil map of Wales that accompanies that publication as an aid to reading and appreciating this book. The map can be accessed freely on the web, by looking at www.landis.org.uk/soilscapes.

I have spent many years in Wales, and much of the time I have been out of doors – either mapping the soils, using the land as an outdoor classroom, or simply enjoying the splendours of the countryside for my own pleasure. In the course of looking at soil, vegetation and landscape in various parts of the countryside, I will suggest some excursions that you might take to experience the land at first hand. These excursions are biased towards the west coast, specifically Ceredigion and Meirionnydd where I have been leading excursions for the past thirty years – I apologise to those who live in other parts of the country, but I have provided plenty of examples and suggestions from all over Wales, so that you can plan similar field trips in your own locality.

CHAPTER 1

Landscapes and Soils

The Welsh landscape is rightly celebrated for its beauty and variety. The fascination of our amazingly diverse countryside is augmented by its associations with a rich culture of literature, history and legend. The land is resonant with the deeds and myths of the past, which are intricately interwoven with the geology, climate, topography and vegetation of the region. Welsh place names are like a list of these cultural influences: some of them reflect the nature of the land, its shape and vegetation, in the earliest days of settlement. Landforms and natural vegetation are the raw materials which have been used by human beings over the millennia to mould our country into a comfortable living habitat. Nowadays most people dwell in lowland valleys or on the coastal plains, and the intricate, mundane paraphernalia of modern life have obliterated much of what went on in the past, whilst the scattered farms and open lands of the uplands still have a bounty of fascinating archaeological remains – tantalising shadows and traces of people's lives from millennia ago. Both lowlands and uplands have been used for homes and fields, the uplands for extensive grazing and for a variety of uses including forestry, mining, quarrying and hunting. Different rocks and stones have been used in different ways to construct dwellings, barns, burial mounds and walls. Characteristic hedge banks or *cloddiau* separate the fields, and their design varies across the country according to the stones and earth and plants available for their construction. In many areas the stony fields have been cleared, and piles of stones are scattered across the landscape. Some of these piles turn out to be ancient burial mounds, like those at the remote, but magnificent Carneddau Hengwm, north of Barmouth, whilst dolmens, cairns and tumuli and hut-circles mark the passage of prehistoric men and women.

It is true to say that the natural environment has moulded the culture such that the Celtic tribes of western Europe developed in different ways from the forest-dwelling Germanic tribes of the east, and the varied peoples of the sunny Mediterranean with its rich variety of once exotic resources. Now of course we take it for granted that we can buy olive oil, rice or wine from the local supermarket.

Below I will try to summarise the main elements of Welsh environments:

An **oceanic climate** with cool summers and mild winters. This may seem a rather bland

Cloudscapes over Ynys Lochtyn, Ceredigion

statement to anyone who has spent chilly, rain-soaked days on the Welsh hills, but it does reflect the fact that you can as easily be cold and wet in the middle of summer as you can on New Year's Eve! The proximity of the ocean and the estuaries surrounding us on three sides has the effect of moderating temperatures throughout the country, but especially on the most westerly peninsulas. The water acts as a kind of storage radiator, absorbing heat during the summer, and slowly radiating it in the winter. This effect is so marked that the seasons are somewhat delayed, making August the warmest month, and February the coolest in western parts like Anglesey, Llŷn and Pembrokeshire. These are the traditional areas where the general lack of hard frosts makes early potatoes a profitable crop, whilst certain plants, like the exquisite, jewel-like vernal squill, occur only in these most oceanic regions.

The climate is certainly wet, especially in popular walking areas like the Brecon Beacons and

Vernal squills on the Pembrokeshire coast

Snowdonia. The highest rainfall is found at high altitudes in the west, and it is worth remembering that the temperature falls by about 0.6 degrees per 100 m rise in altitude, so a mild day at sea level may turn into something quite savage at 800 m. This is especially so because the climate is both variable and generally windy. If the weather turns colder, wetter and windier as you climb, you can quickly find yourself in trouble unless you are well prepared. The well-known wind-chill factor is important here since the higher mountain slopes are much windier as well as cooler than the valleys.

In general rainfall is lower on the eastern slopes of the hills, where a rain shadow effect is in operation. Moist air from the Atlantic rises and cools when it hits our western hills, and drops large proportions of its moisture – leaving less to fall in the rain shadow. The complex topography also engenders variations in climate known as microclimates. Mountain walkers often notice sharp changes in shelter and temperature depending on wind direction and gustiness, whilst north-facing slopes are significantly colder than south-facing ones.

The warmest, coldest and wettest climates of England and Wales all occur in Wales, reflecting the large altitudinal range – from sea level to nearly 1,100 m (3,600 ft) on Snowdon. In fact, the bioclimatic map of England and Wales, published by the then Soil Survey of England and Wales in 1980, records a unique climate in the vicinity of Ammanford, up the river Loughor from Llanelli. Here we find land in the warmest category in the classification, but also in the wettest – a veritable tropical rainforest. By the way, I have stopped thinking of Wales as a wet country since visiting the South Island of New Zealand, where the south-west coast has three times the rainfall of the Brecon Beacons!

A complex **geology** comprises practically every age in the stratigraphic column. Rolling, elevated moorlands of siltstone and mudstone dominate, with some notable exceptions, including the highest mountains of Snowdonia which are formed from a variety of igneous (volcanic) rocks, and the Preseli Mountains of Pembrokeshire, which are also strongly affected by igneous rocks. Ancient sandstones form the bulk of the Cambrian Mountains, the Rhinog Mountains and the Clwydian Hills, whilst red sandstones in Breconshire, Carmarthenshire and Pembrokeshire give mountainous country in the Black Mountains, the Brecon Beacons and the Bannau Sir Gâr (known locally as Y Mynydd Du).

The south Wales coalfield occupies some 2,000 sq. km of mainly elevated moorland between Abertillery in the east and Kidwelly in the west: north-south it stretches from Brynmawr to Pencoed, and it extends westwards across Carmarthen Bay into parts of Pembrokeshire. The coalfield is surrounded by a narrow ring of limestones which reach the coast in south Pembrokeshire and on the Gower Peninsula. Limestones are also present in parts of Anglesey, on the Great and Little Ormes (near Llandudno), and in strips either side of the Clwydian Hills, with a further narrow belt running north-south along the western boundary of the north Wales coalfield, west of Wrexham and Oswestry in the vicinity of Llangollen.

This intricate geological framework has produced large areas of upland and mountains, the result of major earth movements in which, time after time, Britain has been crushed and buckled in the collision of continental plates carrying the land masses of Europe, Africa and North America.

The **topography** is amazingly complex for such a small country, and is dominated by hilly country with steep cwms and deeply cut rivers and streams. It has been moulded by the action of glaciers over most of the land. Ice, over the last two million or so years, has reworked the surface,

Castell y Gwynt, 1,000 m up on the Glyders, Gwynedd

creating U-shaped valleys in the mountain zones, and coating much of the land in a thick layer of heavy boulder clay. Melting ice at the end of the Ice Age has carved deep gorges in many places, and deposited swathes of gravel in others. The tops of the higher mountains are shattered by frost and, in Snowdonia in particular, crazy jumbles of rocks have engendered many place-names from their resemblance to castles and forts.

Land use reflects these conditions of climate, geology and topography. Sheep-grazed pasture predominates in all of the lowland zone, with arable fields mainly confined to the east of the country. Some parts, like the valley of the Tywi and the coastal plain of Ceredigion are famed for dairy farming, as the combination of soils and climate means that good quality grass can be grown. In contrast, many parts of our Welsh land are of limited use for agriculture, and extensive upland conifer forests were planted in the years following the two world wars. Wild habitats abound – from the extensive cliff, dune and other coastal habitats to wet pastures known as *rhos* in lowland, *ffridd* grasslands on mid-slopes, oak woodlands, especially in steep valleys, and an assortment of moorland and mountain vegetation types. Limestone vegetation is not extensive, but is particularly important because so many unusual species occur.

Cadair Idris from Rhydymain

There is just one factor that incorporates all of these environmental influences, and that is the soil. The type of soil is determined by all of the above factors – by climate, geology, topography, land use and vegetation. It is thus an integrating factor, especially since the relationships are complicated and dynamic. Soil determines what kinds of vegetation are possible, but the vegetation also modifies the soil. Soil and vegetation can even affect climate: mature forests, for example, have a more humid, cooler microclimate than open grasslands in the same region.

What *is* soil?

Different people mean different things when they use the word soil. To civil engineers it is loose surface material that isn't rock, so geological deposits like estuary mud, lake clays, boulder clay and sand dunes might be included. To gardeners and farmers it is, perhaps, the top 40 cm or so of the ground, the bit that you dig, plough and sink fence-posts into. A grave digger is traditionally interested in the top 2 metres! For the purposes of this book, we are close to the grave digger. Soil scientists and surveyors who map soil types study the surface of the earth to about 1.2 metres. In practice, many Welsh soils are quite shallow, as they are over hard rocks, but we need a cut-off point, as you can keep on and on digging in some sandy and silty soils. A depth of 1.2 metres is a practical depth, as the health and safety legislation insists on shoring and hard hats once you dig deeper than this. Much survey work is done with a hand auger – a screw tool like a wood auger that brings up samples. Soil augers are usually one to 1.2 metres long. (I have sometimes seen this tool confused with an *augur*, something of a less scientific nature, but also useful in predicting which soils may be found where.) So, for the purposes of this book, soil is defined as the physically altered top 1.2 metres of the earth's crust. We will see that physical and chemical alterations of the bedrock or original parent material are very significant, especially for the growth of plants.

Ploughed land near Aberystwyth

We can all recognise **topsoil** when we see it. It is a dark, crumbly looking material that readily breaks up in the hand, and leaves your fingers dirty. Most people, if asked, will say that it is a mixture of mineral particles, like sand and clay, and material derived from decomposed plants – which is a pretty fair description. Moist topsoil shaken in the hand will break up into lumps of various sizes, but these lumps are obviously made of some kinds of mixed materials stuck together, and will break down further if we grind them between our fingers, or with a pestle and mortar. Those who study soil have agreed to consider the basic mineral particles in four categories, depending on their size. The finest particles are known as **clay** defined as having a diameter of less than 2 micrometres (µm, a micrometre is one thousandth of a millimetre. Just for comparison, a red blood cell is about 7µm in diameter). Clay does not easily separate into individual particles as it is very cohesive, so to study it we have to disperse it in water using chemicals that break the bonds between particles. The next size of particle is called **silt** and has a diameter of 2 to 60µm. Particles larger than this can be seen with the naked eye, and are known as **sand** sometimes divided into fine, medium and coarse sand. Sand, silt and clay constitute what is known as the fine earth. Particles larger than 2000µm (2 mm) are included with **stones** which vary from fine grit, or gravel to large boulders.

Soil samples, of course, are never composed of just one of these particle sizes. They contain mixtures of sand silt and clay, and this mixture is conventionally called the texture or **particle size class** of the soil. Stones also occur in most soils, but they are considered separately, and do not constitute part of the particle size class. The word loam has come to be used for a good mixture containing enough clay to provide nutrients and enough silt and sand to give good drainage and water-holding properties to the soil. Thus soil scientists refer to sandy loams, silt loams, silty clay loams and clay loams depending on the proportions of these particle sizes in the mix. Soils with more than 35 per cent clay size particles are called clay, whilst those with more than 80 per cent sand are called loamy sands and sands. The majority of Welsh soils are in the clay loam and silty clay loam categories.

The dark colour of topsoil is caused by rotted and broken-down remains of plants and animals, which eventually turn into a very complex material, known as **humus**. The more general term that is used for the solid, non-mineral component of the soil is **organic matter.** This is the stuff that sticks to your hands and makes them dirty. In fact, it sometimes reacts chemically with fluids on and in the skin, staining the surface cells. (Soil scientists know this well, as it takes ages to clean hands that have been dealing with humus-rich soils.) The organic matter in the soil is often chemically bound to clay minerals, or coats the surfaces of larger sand and silt particles. We are also likely to see some recognisable fragments of plants – pieces of stem, leaf, flower, fruit, root etc. in various states of preservation, and often some animals like earthworms, ants, millipedes, slugs or insect larvae. A magnifying hand lens will reveal more detail, including ways in which the humus is distributed through the other materials, with smaller soil fauna like springtails and nematode worms. The architecture of the soil becomes evident, with living and dead roots and fungal strands often forming dense mats near the surface, and leaving impressions in the form of pores. The small animals also create pores, whilst the casts, or droppings of earthworms and other animals may be visible. Sometimes, there are chemical concretions that have precipitated out in the soil; small nodules of manganese and iron for example, or larger fragments of iron pan – we will discuss this later in relation to waterlogged and upland soils.

We begin to get the feeling that this is pretty special and complex stuff. That feeling is right!

Sandy soils on dunes, south Wales

Chemists have spent centuries prying into the chemical components of soil, and are still some way from identifying all the molecules that make it up. This is not so surprising when we consider the origin of the materials in soil. The minerals, which is to say the non-organic parts, are derived from geological deposits that were there before soil formed. These, in themselves, are chemically and physically complicated, having been created originally from the earth's mantle in volcanic activity that heated and pressurised the rocks near the earth's surface. Although clay particles are minute, their molecules are very complex and varied, depending on the chemical origin of the rocks from which they are derived. As well as being chemically versatile, the molecules in clay have complicated three-dimensional structures that affect the ways in which chemical elements attach to them, and they play an important role in the storage and distribution of important plant nutrients. They also affect the acidity or alkalinity of the soil. This important characteristic has a fundamental effect on what will grow there. The rocks from which the soil particles (sand, silt, clay and stones) derive may be acidic in themselves. Many of the volcanic rocks and sandstones and siltstones of Wales are in this category. However, there are parts of Wales, notably around the outside of the southern coalfield, but also in Flintshire, Pembrokeshire, Anglesey and around the Great Orme where limestones are important. These have distinctive soils and vegetation that we will describe later.

The **organic soil constituents** are even more complex than the minerals, consisting of the breakdown products of plants and fungi mixed with substances excreted by animals and micro-organisms. Some of this material is readily recognisable as fragments of plant tissue, but most of it is broken down – either by passing through the digestive tracts of animals, or by being incorporated into microbial material. This **humus** is found in particular parts of the soil – especially in the topsoil, where it glues fragments of soil together and coats the surfaces of the resulting structural units, which are known as **peds**. If you shake some garden soil in your hand, it will usually break into breadcrumb-shaped fragments a few mm to a couple of cm in diameter. These are the peds. Their surfaces are often coated with humus, whilst pores that run through them may also contain enhanced amounts of this material.

This leads us on to the physical **structure** of soil. Peds are a unique feature of the material, and can be used to identify ancient soil deposits, for example, below the archaeological occupation layers formed from the debris of human lives, the remains of buildings and the accumulated junk that people have discarded over the centuries. At the base of these layers, it is usual to find soil that has been preserved by the overlying layers. Below mediaeval and Roman layers in York, for example, I have seen topsoils with peds and the remains of earthworm casts still identifiable after 2,000 years.

Soil is never homogeneous – the solid mineral parts are mixed with humus, air and water to form a complex system in which the 'whole is more than the sum of the parts'. The mixture of mineral particles giving the texture is itself organised into the larger, compound peds – soil structural units. Subsoils often form into blocky or prismatic peds, which may have angular or more rounded surfaces. These form by the wetting and drying of the soil, by the binding and secreting actions of roots and by the activities of the soil animals. They often have a zonation of organic matter content, with most of it coating the surfaces of the peds. The ped surfaces may be coated with other materials such as iron and manganese oxides, or clay that has been redistributed by water flowing through cracks and pores and over the ped surfaces.

Typical soil profile (Brown earth) in grassland, Meirionnydd
Note the dark A horizon, with a brown B horizon below it passing into increasingly stony, shattered bedrock

Porosity is a subdivision of structure: pores and fissures occur between peds, but also within the ped structures. They form as a result of root penetration and decay, and through the activities of small boring animals like nematodes and pot worms. Larger earthworm channels extend from the surface to as much as 1.5 m in depth. I'm sure that you realise that I don't mean boring in the sense of tedious here!

So far I have been talking mainly about the **topsoil**. This is the dark brown, black or grey layer extending about 20 to 30 cm below the surface. Under wild (semi-natural) vegetation it contains a dense mat of roots from the surface to about 10 cm down. In grassland it is this root mat that can be scraped up as turf and used for laying lawns.

Many people regard the topsoil layer as *the* soil, and don't think too much about what lies below it. However, if you look at soil sections exposed in road sides, quarries and so on (or go to the trouble of digging a hole), you will notice that there is a distinctive layer immediately below the topsoil, often called the subsoil. It is not so dark because there is less organic matter, but it still has several characteristics which distinguish it from the rock or the other material below. It is often a brownish colour, whilst closer inspection will show that it still contains roots and soil animals, and the pores, fissures and peds that we saw in the topsoil. These are all soil characteristics and can be used to differentiate soil from non soil.

The material from which the soil has developed is called the **parent material** (see box). It is conventional to call the humus-enriched topsoil the **A horizon**, whilst the distinctive layer below, which is not as dark and has fewer roots is called the **B horizon**. The B horizon is of particular importance as its depth, structure and wetness are critical to plant growth, and to the way the land can be managed. The colour and structure of the horizons below the topsoil give us vital information about the character and potential of the soil, and are used to classify the soil. The relatively unaltered parent material below this is called the **C horizon**. The C horizon may be solid rock, much-weathered rock, or a variety of loose deposits like sands, silts, clays etc. In Wales the rock surfaces were often strongly altered by freezing and thawing during the period just after the glaciers retreated (about 12,000 years ago). During this period too, large swathes of boulder clay were deposited from the glaciers leaving a material (properly called till) that contains chaotic mixtures of sand, silt, clay and stones, sometimes including large boulders.

So, soils have these distinctive layers, called **horizons**, and the nature and distribution of the layers can be used to classify different kinds of soils. There are many different kinds of soil classification in different parts of the world. In the next Box, I give a summary of the classification used in England and Wales by the National Soil Resources Institute.

Soil classifications enable us to map the distribution of soils, to think about how they can be used, and the ways in which they determine vegetation and landscape. The classification of soils has been used extensively all over the world in order to understand the characteristics and distribution of different soils, to improve productivity in agriculture and forestry and to help ensure wise and sustainable use of land. In the following accounts of the soils, landscapes and vegetation of Wales, I will refer to something called the **Soil Series**. This is the equivalent of the *species* in biological classifications. The Soil Series are what are usually shown on detailed soil maps, and are given names, usually referring to the places where they were first mapped. Many of these names are Welsh – as Wales was in the forefront of mapping British soils.

Soil parent materials

In a wistful allusion to genetics (which gives such a useful basis for classifying living things), the raw materials from which soils form are called **parent materials**.

The formation of a new soil depends on the provision of new potential soil material – either by denudation – the scraping clean of a surface by the action of wind, water or ice – or by deposition of new materials from erosion elsewhere – for example river gravels, rock falls or blowing sand – or the formation of new rocks by uplift of the ocean bed or volcanic action.

There is thus an almost infinite variety of possible parent materials, all with unique characteristics which will be incorporated into the soils formed from them.

Here are some examples from Wales:

- Rocks – sandstone, siltstone, mudstone, limestone, volcanics
- Gravels – river and sea deposits
- Scree – fallen rocks
- Alluvium – silts, clays from marine, estuarine or freshwater rivers and lakes
- Boulder clay – can be very variable deposits laid down by scraping ice
- Loess – blown silts and clays are common inland in continental Europe
- Blown sand – forming dunes or plains
- Man-made materials – e.g. slate or coal mining waste

Soils form from parent materials by processes of **weathering** – i.e. the breakdown of the material under the influence of wind and water, accelerated by the chemical and physical effects of plant roots and animal action. Soils are also formed by the formation and decay of organic matter and by **pedogenic processes** – the leaching effects of water and the biochemical actions of bacteria and other living things.

Soil mapping involves identifying the types of soil present in an area and drawing lines on the map to separate them from each other. Sounds simple enough, doesn't it? But there are a number of snags with this. If we could dig a 1.2 m deep hole every 10 m across the countryside and identify every soil profile exposed, we could then draw lines on the map between the sites of the profiles, separating groups of profiles that were similar. However, a moment's reflection on this shows the problem. Take the soil map of Wales as an example. The area of Wales is about 20,000 square km. If we dig profiles every 10 m, we end up with 10,000 holes per square km – or 200 million for the whole of Wales! If it took a modest 30 minutes to dig and identify each profile, the exercise would take 100 million hours – well over 34,000 years of eight-hour days! So we have to drastically cut down on

Manod Mawr, near Blaenau Ffestiniog gives its name to the Manod series, a well-drained soil of the ffridd lands – most of the soils in the foreground of the picture are Manod series

the number of profiles that we examine. In fact, it would be pointless digging holes at such a frequency in many areas where the soil varies little from place to place. The flat lands around estuaries for example, often vary very little within short distances. What the surveyor does is to create a landscape model, or hypothesis, using previous experience, and other evidence from the landscape, and then test this with trial profiles in suitable places. In the example of the estuarine landscape one may suppose that all profiles at a given altitude just above sea level will have very similar soils. Small elevated areas may be supposed to be levees (see p.31), whilst sunken areas may represent old river channels that have silted up. The surveyor will test these areas to see if this hypothesis holds. More complex landscapes will require more sophisticated hypotheses, perhaps aided by the use of geology maps, aerial photographs etc.

So, that's one problem dealt with. But there are others. The classification is very precise. It is unlikely

How Soils are classified in England and Wales

Soil types that can be mapped as essentially consistent types across the landscape are called (rather confusingly!) ***soil series.*** Each soil series has lots of characteristics that define it, and to enable us to handle the large numbers of soil types, they are put into major groups, each of which has one or more defining characteristics. For example, **Brown soils** are soils with brown or reddish, un-mottled B horizons, and are more than 30 cm deep. The major groups are subdivided into soil groups; for example, within the Brown soils is a group called Brown alluvial soils, defined as being developed in loamy or clayey alluvium more than 30 cm thick.

There is no need to worry about the details of this classification, but here they are anyway! Some readers may like to put the soils of Wales into their national context, whilst the classification presented here, in a simplified form, will help in the interpretation of published soil maps in the UK.

Major Gp. No.	Name	Area (% Wales)	Description
1	Terrestrial raw soils	<0.01	In recently formed well-drained material with little or no soil formation.
2	Raw gley soils	0.21	In recently formed waterlogged material with little or no soil formation.
3	Lithomorphic soils	2.21	Very shallow (<30 cm) soils over rock or little-altered loose materials.
4	Pelosols	0.10	Relatively well-drained heavy clay soils with strong structures.
5	Brown soils	30.22	Deep, well-drained permeable soils.
6	Podzolic soils	32.30	Soils with a peaty or humose surface layer and subsoils enriched with iron or organic carbon.
7	Stagnogley soils	24.63	Slowly permeable, seasonally waterlogged soils with perched water table.
8	Groundwater gley soils	3.26	Permeable, seasonally waterlogged soils affected by a groundwater table.
9	Man-made soils	0.43	Disturbed, restored or man-made soils.
10	Peat soils	3.39	Soils with more than 40 cm of organic material.

(sources: Avery, 1980; Rudeforth et al., 1984, adapted from Conway http://www.rac.ac.uk/?_id=3724)

that the landscape units identified for the hypothesis will contain 100 per cent profiles of one soil type. This means that we have to generalise about the contents of the areas drawn out on the map. So when an area of alluvial plain is separated out as, for example, the Newchurch series, this should mean that digging in that area is more likely to reveal a soil conforming to the definition of Newchurch than any other soil series. If it is not Newchurch it should be something very similar. The definition of Newchurch soils includes "stoneless, clayey, mottled, with calcareous horizons above 40 cm". If the profile that you dig within an area mapped as Newchurch is stoneless, clayey and mottled, but non-calcareous you shouldn't be too disappointed – at least the map is three-quarters right! What is more, the soil is still likely to have a relatively high lime content, and surrounding profiles are likely to be calcareous. In many parts of Wales, especially on till (boulder clay), the soils are rather variable in their particle size class, but in most cases it will not matter too much whether you have, for example, a Brickfield series profile which is dominated by clay loam, or a Cegin profile dominated by silty clay loam, as these soils are identical in all other respects.

These considerations may seem rather arcane, but they have practical implications. It is not wise, for example to take a small scale soil map like *Soils of England and Wales sheet 2 Wales* at 1:250,000 scale (4 cm = 10 km, or 1 inch = 4 miles) and blow it up to a much larger scale. This may seem obvious, but I have seen it done more than once! As well as exaggerating boundary errors, this will give a false impression of precision. In fact, the national maps do not pretend to map out a single soil series, but associations, with one soil series usually (but not always) dominant, and a range of other soil series that may occur within the units. The information given in this book will help you to sort out the likely variations of soil that you will find within the associations mapped.

Like most disciplines, the study of soil is bedevilled with many technical, but necessary specialist terms. No doubt you have already become aware of that fact in this section. I have tried to define these as we come across them, but the browser may find him/herself tripped up from time to time. To help out here there is a glossary in the back of the book.

CHAPTER 2

Estuaries

The extensive estuaries of Wales have a very special and distinctive charm, giving a recognisable *genius loci* – or spirit of place. From Malltraeth on Anglesey to the Usk and the mighty Severn in the south east we are never far away from these haunting places of wide skies, mudflats and waving reed beds. These remotely beautiful and lonely places have inspired many artists, writers, poets and photographers. Their special character is epitomised in a descriptive passage in *Owen Glendower*, by John Cowper Powys, where the army of Glyndŵr is marching along the Mawddach below the slopes of Cadair Idris. He comments on the strange, almost alien atmosphere of this long arm of salt water stretching far inland.

There are several stunningly beautiful estuaries around the Welsh coast – at Conwy, Porthmadog, Llanbedr, Barmouth, Aberdyfi and Cardigan to name a few, each having wide areas of semi-natural vegetation with a variety of colourful scenery and ideal habitats for many distinctive animals and plants. The scenic qualities of these areas spring from the combination of wide open areas of water, sand or mudflats, and the rich contrasts with the forests, fields and moorlands of the surrounding hills. The Mawddach for example, has glorious prospects of Cadair Idris whilst the Conway and the Dwyryd are backed by the mountains of Snowdonia. Two large estuaries help to define the very shape of Wales, those of the Dee and the Severn.

Flowing water always carries a burden of solid particles. If you have a bath after a day on the beach, you will notice the deposit of sand that lies in the bottom of the tub after you have let the water out. This is a **sediment**. Rivers carry the solids suspended in the water, and the size of the particles carried depends on the rate of flow. If the flow slows down, then the water is unable to carry coarser sediments, and they drop out onto the bed of the river. This is why your bath gets dirty: if you could keep the water moving fast, the particles of sand would all be washed down the drain, but the drainage slows down as the last dregs of water disappear, depositing the sand around the plug hole. Deposits like this are called **alluvium**, and the land on the flat areas around estuaries and rivers is formed from alluvial deposits.

Two of the most interesting habitats are the lower **salt marsh** – the area of flat mud exposed at mid and low tides, and carrying sparse, but more or less natural vegetation, and the upper salt marsh,

The Mawddach estuary

sometimes called saltings – semi-natural grasslands which are inundated by the highest tides, but provide good grazing. The admirable tradition for local specialities and good food in France means that its saltings are particularly prized for the distinctive taste of *agneau de pré salé*, saltings lamb: this speciality is being developed here in Wales too, where the valuation of good local produce is going from strength to strength. We also have some other tasty traditional specialities, including cockles from the Penclawdd salt marshes, glasswort, sometimes known as samphire, or sea asparagus, and laver bread, a seaweed delicacy collected from the rocky shores of south Wales.

These estuarine habitats serve to underline the difference between soil and sediment. One of the definitions of soil is *the material in which plants grow*. The sparse vegetation of the salt marsh grows in an unconsolidated and undifferentiated silty deposit. It has been laid down by the slow-flowing waters of the river as they are held up by tides and flood over the areas at the tidal limit. The sediment is almost constantly waterlogged as the fine silty particles do not drain quickly enough to dry out between tides. It is rich in organic particles – but these have been carried in the sediment, and have not developed in the way that soil organic matter usually does, by the rotting of plants in situ. If you dig into this material (make sure you won't sink into it, or get cut off by the tide!) you will soon experience the strong bad eggs smell caused by sulphur bacteria – they operate under waterlogged conditions that are generally unfavourable for soil formation. This stuff is a parent material – a mineral and, in this case, partly organic material that has the potential for forming a soil – but has not yet done so. It *is* layered, however, having a thin brownish surface layer that is rich in iron and hydroxide, and a gloopy, smelly lower horizon that is black with iron sulphide.

In the most frequently inundated parts of the salt marsh, only one flowering plant may occur – the cord-grass *Spartina anglica* (see Box p.32). Other plants of the lower salt marsh include eelgrasses and saltworts. Nearer to the shores, an increasing number of highly specialised species grow, including salt marsh grass (*Puccinellia maritima*), sea arrowgrass, and the colourful flowers of thrift, sea milkwort, sea aster, sea spurrey and sea lavender.

Sedimentation

As rivers open out approaching the sea, so the energy of their waters is dissipated over an increasing area. At the same time, the incoming tide slows down the flow, and there is regular flooding with the spring tides. This widens the inundated area even more. As we have seen, the amount of sediment carried by the flowing water depends on its velocity, or speed of flow. Rushing water will move large boulders, as can be seen in several of our narrower river valleys (see chapter 6 on Ceunentydd). In major estuaries like the Dee, Dyfi, Mawddach and Severn, silt and fine sand are deposited by the slow-moving waters. As the main

Salt marshes on the Dwyryd estuary, with Portmeirion in the background

Undeveloped soil below the high tide mark, in Severn estuary

Alluvial gley soil (Newchurch series) in Severn estuary

channels overflow with tidal movements, slightly coarser sediments form ridges as they are deposited, this creates a natural bank along the course of the creeks, termed a **levee**. These can block the flow, causing the channels to re-cut and change course. During floods silt is deposited above the levels of medium tides, and eventually becomes a soil – a substrate for the growth of rooted plants like *Spartina*, the cord-grass (see Box p.32). Once vegetation becomes established, it traps even more sediment – an example of positive feedback – and the land begins to reclaim the estuary as salt marsh. The situation usually remains unstable however, as storms and big tides will undo the work of sedimentation from time to time, and begin a period of erosion. The eroded material will be deposited

The **history of the cord-grass** is interesting as it is a good example of ecological and genetic adaptation and colonisation. The cord-grasses have a special method of carbon assimilation which makes them more competitive in situations of high light intensity. In open estuarine habitats they can take advantage of the relative lack of competition from other plants, whilst sunlight reflected onto the undersides of the leaves confers a bonus of added light intensity.

In the early nineteenth century an American species of the grass (*Spartina alterniflora*) was introduced in Hampshire. By 1870 it had hybridised with the native cord-grass (*Spartina maritima*) to form a vigorous but sterile plant with a double set of chromosomes. This plant (called *Spartina x townsendii*) was able to spread very successfully by vegetative methods. In 1892 a fertile hybrid was found in Southampton Water. The incompatible chromosome sets from the original parents had doubled up, and become compatible so that sexual reproduction could occur. The new species was given a new name – *Spartina anglica*. It was so competitive and successful that it quickly spread around the British coasts, and around the world. It is now the dominant plant on most of the Welsh salt marshes. The rapid spread and growth of the plant has increased sedimentation rates and caused economic problems by silting up shipping lanes, as well as triggering ecological changes comparable to those wrought by *Rhododendron* in the native woodlands. We should remember that living things do more than adapt to their environment – they also modify their environment.

somewhere else, and the cycle restarts. On most estuaries people have taken a hand in the process, and built embankments that minimise, or control flooding. This allows mature soils to form, and it is sometimes possible to relate the degree of soil development to the age of reclamation from the sea.

I have already described the sulphurous proto-soil of the salt marsh: inside the protective sea wall, soils have had time to develop over at least 500 years. Here there are soils known as **gley** soils – from a Russian word meaning clay, although gley soils are not necessarily clayey. Gley soils are always prone to waterlogging, and are much wetter than Brown soils. They are distinguished by generally greyish colours and scattered mottles (fine blotches, like a rash) of reddish and orange oxides of iron.

The gley soils in estuaries are known as **alluvial gleys**, and are usually characterised by a lack of stones. Those around the Welsh coast have heavy textures, with clay and silt predominating. The National Soil Map shows that most of our estuarine soils are mainly silty, mapped as **Tanvats Association**. The soils on Morfa Rhuddlan in north Wales and on the Gwent Levels in the south-east are more clayey, Wallasea and Newchurch Associations respectively (see photo above). The land is used for grassland, often for

Gley colouration is one of the most important characteristics of soils, since it can be used to indicate how wet they normally are. Freely drained soils – those that are hardly ever waterlogged – are of a uniform brownish colour below the dark topsoil. It is the iron in soil that gives it its brown colour, and when the soil is well aerated, the iron is predominantly in the oxidised form – like rust. In soils that experience some waterlogging, there is a tendency for the iron compounds to be in a reduced form, with less oxygen in the molecule. These compounds are greyish, so the background colour of the soil is grey but, as there is some aeration, often around cracks in the soil, or where roots have penetrated, flecks of brighter-coloured iron oxides, known as mottles, are found. The absence or presence of these colours is an essential diagnostic character when assigning soil profiles to one of the classes in the classification.

Peaty gley soil exposed in forest ditch, mid Wales

grazing cows and sheep, but farmers must keep a vigilant eye on tidal conditions and weather forecasts to ensure that their stock do not become trapped by floods. The surface of these soils very easily becomes puddled and messy if livestock is allowed on them when the weather is wet. In fact, it is best to remove animals from the land as soon as the warmer, drier summer period is over. In practice, there is a balance between the water coming into the land via rain, and that which leaves the soil through evaporation and transpiration. In these alluvial soils there can be additional water rising from below as groundwater. The groundwater in tidal zones responds to the ebb and flow of the sea or estuarine water. Much of the estuarine pasture country has been drained by pipes leading into the ditches, but in some areas the groundwater is too high for this to be very effective. The ditches too are special, as precautions have to be taken to prevent high tidal waters from backing up the ditches and flooding the land.

If you examine a section, or profile, of the soil you will see a dark stoneless topsoil, with some yellow-orange mottles. Permanent pastures have shallower, but more humus-rich tops, and a dense root mat. Below this the B horizon is stoneless and decidedly light grey in colour, with a few reddish-brown and orange mottles. An excavated soil profile usually shows cracking where the soil has dried out during dry summers, and roots and earthworm channels favour these cracks. If you prise out the lumps of subsoil you will find that the peds are coarse, 10–20 cm in diameter, and prismatic in shape, although they may break up into smaller, angular units.

This type of land is the result of reclamation projects which have been going on since at least Roman times. The fields were, and often still are, liable to flooding at high spring tides, or when winds pile up the tide as it advances up the estuaries. It has thus remained pastoral and without houses – frequently remote and attractive, with a good range of wetland habitats in the areas too wet for agriculture. Fields are separated by banks and hedges with pollarded willows and alders; whilst willow carr vegetation – wet woodlands with willow and alder – can be found in the wetter zones, which may have developed peaty topsoils. There are even deep peats in some areas, like parts of the Gwent Levels, and in the Mawddach and Dysynni estuaries. These wettest areas are often managed as nature reserves. The Teifi marshes upriver from Cardigan, for example, are managed as a large nature reserve and Site of Special Scientific Interest, and house the headquarters of the local Wildlife Trust. The larger areas are important feeding areas for birds, and their remoteness and inaccessibility favour shy animals like the otter.

Some exciting finds have been made in the mud of estuaries, including ancient boats which have gradually been buried in silt over the millennia, but I don't recommend that you go mud-hopping to look for them! An easier way of exploring the past is to look for prehistoric monuments on the drier, more hilly lands adjacent to the flat lands. The standing stones, Meini Llwydian on the Taf near Llansteffan, Carmarthen, and the burial chamber at Hendre Waelod, Bryniau near Tal-y-cafn, Conwy are good examples. At Llanbedr, Gwynedd, standing stones mark the route of a prehistoric trackway leading from the old harbour north of Shell Island. Nearby, at Llandanwg, the ancient church is periodically buried by sand dunes.

Excursion – The Mawddach Estuary

OS map Explorer or Outdoor Leisure 23 – Cadair Idris & Bala Lake / Llyn Tegid (SH629141)
Total distance: a little less than 3 miles, mostly flat and easy.

Start from the platform of the old **Morfa Mawddach station**, now a car park, 2 km north-east of the holiday village of Fairbourne. Go north on a footpath across the salt marsh, parallel to the railway line, to Fegla Fawr – the larger of two rocky hillocks which were islands in the estuary before the surrounding mudflats silted up. There are fabulous views of the Cadair Idris range from here, and you can enjoy the bracing smell of the salt air. Follow a footpath that climbs over the grasslands and oak woodlands of the 'island' to arrive at Mawddach Crescent. This unusually positioned crescent of tall terraced houses was built at the turn of the century to attract visitors to the area following the opening of the railway. A footpath from here skirts the north side of the 'island' where some of the silting up processes can be observed in the wide estuary. The path then goes to the south of a second 'island' – Fegla Fach. The screen of Corsican pines on this smaller rocky hill conceals a swish residence built by the owners of the eponymous curtain company. Between the two 'islands' a track goes south-east to the main road, passing an area of raised bog, with open mire vegetation, willow scrub and patches of bog myrtle. There is a considerable area of peat in the zone protected by the islands, which has been above the high tide level for at least a thousand years, giving time for deep peat to develop in the waterlogged conditions. This is not an unusual juxtaposition in estuaries, and there are extensive peat deposits adjacent to many of our Welsh salt marshes.

Continue south along the bank of the tidal river Arthog, and then return along the old railway (closed in 1965, following the Beeching cuts) which is now the Penmaenpool–Morfa Mawddach Walk. The wooded area close to the car park is a reserve of the RSPB – Arthog bog – and has a range of habitats including a raised bog and willow fen or carr. A pleasant path makes a circuit around the reserve if you want to extend your walk a little.

This walk, like many others along the estuaries of Wales, will reward you with glorious views, and an interesting insight into the way that soils and landscape interact. You can do similar walks in many parts of Wales. There are suitable footpaths along parts of most estuaries, although care must be taken to avoid being cut off by exceptionally high tides. Here are some examples:

- Along the river **Dee** between Flint and Bagillt, about 1.5 miles one way (start at Flint Marsh SJ235740, or Bagillt SJ222754).
- Along the river **Clwyd** between Rhuddlan and Kinmel Bay, about 2.5 miles one way (there are rights of way on both sides of the estuary, the west side overlooks Morfa Rhuddlan. This extensive marsh – *morfa* or sea-place – was the scene of a great battle which has found a permanent niche in Welsh folk memory – the

lament of *Morfa Rhuddlan* is one of the oldest surviving pieces of Welsh music. The Welsh force was defeated by the Saxon armies of the famous King Offa in 796. According to John Davies in his *A History of Wales*, Robert Graves called it "The Flodden of Wales".

- East of Bangor, on the **Menai Strait**, a coastal path runs for about 4 miles from the east bank of the river Ogwen, opposite Penrhyn Castle at SH615724, to Llanfairfechan. The path can be accessed from the village of Aber, from which a track also leads inland to the spectacular Aber falls.
- A footpath runs for about 2 miles along the western shore of Foryd Bay, four miles SW of **Caernarfon**, from Blythe Farm (SH448576) to Fort Belan.
- From Pont Croesor (SH593413) on the B4410, a footpath crosses the marshlands of the **Glaslyn**, northwards to Glastraeth. Since the construction of the Cob at Porthmadog, this part of the former Glaslyn estuary is no longer tidal. The footpath is sometimes closed in summer as it passes the recent nesting site of rare ospreys.
- On the south side of the **Dwyryd** estuary, a footpath follows the sea wall from Bryn Glas (SH620376) for 2.5 miles to Glan-y-wern on the A496, a mile south of Talsarnau.
- From Pennal, on the A493 you can walk southwards along the river **Dyfi** for about two miles, as far as the bridge west of Dyfi Junction railway station. It is possible to return on a route further inland via Ynys farm and Plas Talgarth hotel. You can also enjoy wonderful views of the Dyfi, Mawddach and Dwyryd estuaries from the comfort of a Cambrian Coast railway carriage, if you take the train from Machynlleth to Penrhyndeudraeth. Remember to make sure that the window is clean, or you may have a frustrating ride!
- From the ruined church at Llandeilo–Abercowyn (SN310130), a mile north of Dylan Thomas's boathouse at Laugharne, but on the opposite bank of the **Taf estuary**, a footpath goes for a mile northwards, passing the standing stones, Meini Llwydion, to Pont-ddu.
- A footpath follows the inland edge of Llanrhidian Marsh, in north **Gower**, from Llanrhidian village westwards for two miles, passing Weobley Castle to Landimore.
- The **Gwent Levels** cover a large area (over 100 sq. km) along the **Severn estuary** from Cardiff to Caldicot. The area has a network of quiet country lanes, and a footpath follows the coast from Porton house (ST390826) near Whitson, 3 miles eastwards to Magor Pill (ST435847).

CHAPTER 3

Sand Dunes

Sand dunes are exciting places for recreation and for study, and are a frequent and popular occurrence along the Welsh coast. You will often find them in the vicinity of the salt marshes that we have just been discussing. They occur on exposed coasts where cliffs are low or absent. At Gwbert near Cardigan for example, an area of dunes has formed on the top of low clay cliffs. At Merthyr Mawr, near Porthcawl in Glamorgan, the dunes are up to 60 m (200 ft) high – among the highest in Europe. Dunes need a source of sand, and this is usually available near estuaries where sandy sediments are constantly deposited by rivers as they meet the sea. Sand bars are exposed at low tide and, as they dry out, the wind picks up the sand and blows it inland.

The picturesque dunes at **Ynyslas**, north of Borth in Ceredigion, are typical of many areas. They cover about 100 hectares, and provide a whole variety of habitats for some unusual, interesting and characteristic plants. They also provide an ideal habitat for people who come in large numbers in summer to enjoy the sandy beaches and the superb natural history of the area. As well as relaxing on the sand, (usually with a windbreak), people fly enormous kites, kitesurf and windsurf. Others walk in the dunes and look at the glorious scenery that surrounds the spot, with views up the river Dyfi to the Cambrian mountains.

On the western, seaward side of the sand hills, there are highly exposed and disturbed habitats, where plants must cope with the continual movements of sand driven inland by the westerly gales. Sand is continually being moved around, building up new dunes in some spots and breaking them down in others. The coarse tussocks of marram grass are ideally adapted to deal with these conditions and seem to actually *like* being buried. Species like restharrow, sand spurge and ragwort cope with these disturbed habitats by seeding quickly and prolifically. The little screw moss (*Tortula ruralis*) grows on these windward dunes too, and forms patches which stabilise the surface. The moss is able to grow up through the sand as it is buried, and can withstand drought. It looks brown and withered in dry weather, but you can amaze your friends by wetting it, when it immediately turns a bright yellow-green!

Dunes at Ynyslas, near Aberystwyth

The soils of the dunes include loose raw sands, with no horizon development, on the seaward side; soils with an A horizon, in which some humus has accumulated, over loose sand, which are called **sand rankers**; and **sandy gley soils** with grey, mottled subsoils affected by groundwater in hollows, called slacks, between the dunes. The raw sands are usually calcareous due to the inclusion of shell fragments in the sand. These carbonates tend to be washed out of the more stable soils of dunes, but in some areas the rankers are still calcareous and are called sand pararendzinas. (A **rendzina** is a shallow calcareous soil. We will meet it again when we talk about limestone landscapes in chapter 10.) The raw sands are permeable, so they hold very little water, and they are colonised almost exclusively by the drought-resistant marram grass with its long water-seeking rhizomes. This hardy grass is able to withstand constant burial by blowing sand – in fact it seems to grow more vigorously when this happens. Its long rhizomes eventually stabilise the sands, allowing some soil development in the form of incorporation of organic matter into the surface. If the surface remains stable, enough organic matter will accumulate to improve the water-holding ability of the soil, and other species will begin to gain a foothold. Eventually, some other grasses and a mixture of herbs will become established, and the soil resembles that on the chalk downlands of southern England. This is why one of the classic New Naturalist series is called *Downs and Dunes* – the soils and the flora have remarkable similarities on these apparently disparate habitats. Several of the plants found on the more stable parts of dunes are particularly striking and attractive. They include several orchids (e.g. pyramidal and bee orchids), the handsome hound's tongue (*Cynoglossum officinale*) and the evening primroses. Smaller gems stud the grassland – bright purple vetches, yellow ladies' bedstraw, mauve patches of thyme, scarlet pimpernel, restharrow, eyebright, yellow wort and the mauve-pink storksbill to name a few.

Morfa Harlech and **Morfa Dyffryn** are two very large sand dune systems in Meirionnydd. They are separated by the Artro Estuary and the built-up areas of Llandanwg. South of the Artro is the popular holiday camping site on Ynys Mochras, better known to generations of

Tortula ruralis

Blowing sand on the seaward side of dunes

Midlands holiday-makers as Shell Island. Visitors approach the 'island' over a tidal causeway, but in fact the peninsula is joined to the mainland by the dunes of Morfa Dyffryn. Both areas are National Nature Reserves with "a range of sand dune communities exhibiting... extensive dune slacks and many nationally scarce plant and invertebrate species". (Countryside Council for Wales web pages.) These sand dunes are very extensive – altogether covering some seven square km. The wildlife is protected by the sheer size and complexity of the dune system – few of the many visitors to the coast will venture deeply into the dunes.

Dune building

The western shores of Wales in particular, are notoriously windy. If you walk in a dune area on a windy day you may be in for an uncomfortable experience as the blowing sand will drive itself into your eyes, not to mention your hair, ears, neck and mouth! Very specialised plants are adapted to the build-up of sand, surviving the continual burial of their shoots by growing rapidly. The main coarse grass of the dunes, **marram grass** (*Ammophila arenaria)*, is largely responsible for the formation and partial stabilisation of the dunes by binding sand grains together with its extensive rhizomes and roots. The sand couch grass *Elytrigia juncea* does the same thing on the seaward edge of the dunes

Sand ranker with buried A horizons at Ynyslas – note long rhizomes of marram grass, exposed by wind erosion

and, what's more, with the help of some other hardy plants like the sea rocket, it actually begins the process of dune building by trapping wind-blown sand. Once a small mound of vegetation or tide-washed debris is formed, there is a tendency for it to trap sand and become a small hill. You might expect that the sand is blown up against the obstruction, to accumulate on the windward side: in fact, because the wind speed drops in the lee of the growing mound, blowing sand from off-shore banks falls out of the air stream, piling up on the leeward side of the obstruction. The plants of these embryo dunes are rightly dubbed pioneers and, like all pioneers, they are resourceful and tough. They are physiologically adapted to put up with the high salt content in the air and soil, and physically adapted to resist the abrasion of the fast-moving sand grains.

Links

On windy days it's best to walk on the inland side of the dunes where conditions are less severe. Here, in early summer, you can revel in a variety of colourful plants that are fairly unusual in Wales and the glory of our coasts. The **links grasslands** of this zone have a range of grassland herbs which benefit from the reasonably high nutrient status and moderately high pH (see Box p.64) of the more stable soils, especially when enough humus has accumulated in stabilised soils for water retention to be improved. Rabbits just love these grasslands: they excavate extensive warrens in the deep, well-drained, stoneless soils. Here on the inland side of the dunes the soils have been stable for long enough to have some topsoil, and the subsoils are stabilised by extensive root systems. The rabbits are able to create tunnels which seldom collapse, as they would do in loose sands. The mixed herbaceous sward furnishes good grazing, which keeps the swards short and often lawn-like. A certain amount of grazing promotes a good variety of plant species, as it encourages the less-vigorous plants which can cope with having their upper leaves eaten. The level of grazing required for maximum biodiversity is an area of very active research: rabbit populations

Bog pimpernel and marsh pennywort in the slacks

fluctuate naturally on dunes, and have considerable effects. Erosion can be accelerated by heavy burrowing, whilst a reduction in rabbit grazing will promote the extension of scrub species like bramble, burnet rose and elder, and a reduction in the number of different species over more stable parts. Some plants, like the vetches, stonecrops and storksbill seem to favour the areas next to rabbit burrows, being able to compete with other plants that can't cope with the soil disturbance. The connection between rabbits and dunes has given rise to the place names Burrows (at Crymlyn and Margam for example) and Warren (at Newborough, Talacre and Stackpole among others).

The name *links* was first applied to undulating land near the shore, from an old English word meaning ridge. The word *lynchet,* meaning prehistoric or mediaeval cultivation terrace, is from the same root. We now associate links with golf courses. The hummocky topography is interesting and challenging for the sport, and the soils are generally well-drained and easy to manage, allowing good greens and fairways to be maintained without undue expense. It is fairly unusual to find dunes without an accompanying golf course.

Slacks

Eroded areas which have been scoured by the wind (called blow-outs) form hollows in the dunes, termed **slacks**. Groundwater is near the surface, and often rises to cause winter flooding. Frequent but intermittent waterlogging of roots requires adaptations to stress, so proportions of stress tolerant plants are increased here. Stress tolerators tend to be components of interesting, often colourful, mixed vegetation, with a good number of species in a small area. This is because highly-productive or competitive plants, which need rich nutrient supplies, are unable to cope with the various types of stress applied by the environment, so they are unable to swamp the soil surface. Marsh pennywort and bog pimpernel are often found among these damp pastures, and you may be lucky enough to see the striking pink flowers of ragged robin. The slacks are often colonised by the creeping willow and marsh pennywort, and taller eared willows and goat willows may become established. Look among the willows for a striking bright red beetle, looking like

Storksbill around rabbit burrows on the dunes

a robust ladybird with no spots. This is a leaf beetle called *Chrysomela populi* that feeds on willow leaves. In midsummer the slacks can become riotous with orchids including the early marsh, heath spotted and common spotted orchids and the marsh helleborine. At this time of year, on a sunny day, there are few better places to be! In some sand dunes in England, Ireland and Scotland, the rare natterjack toad survives in temporary pools. It became extinct in Wales, but has recently been reintroduced to a site at Point of Ayr, where it is doing very well.

Land near dune systems is often affected by an increase in sand blown into the surface horizons. Gley soils on the flat lands around Broadwater, north of Tywyn, Meirionnydd, for example are altered in this way. Sand dunes are too hostile an environment for anything but rough grazing, and nowadays their value as wildlife habitats far outweighs their agricultural value. Some dunes have been afforested, notably those on Newborough Warren in southern Anglesey, parts of the Harlech dunes and at Pembrey, west of Burry Port and Llanelli. Wind exposure is anathema to foresters, but the hardy Corsican pine does well on the more sheltered parts of stable dune systems. I have my own theory that the broken ground of dune systems, as well as providing local areas of relative shelter, can slow down regional airspeeds by causing turbulence. Also, it may be possible to establish wind-hardy trees like the Corsican pine in the lee of protecting shrubs, like the orange-berried sea buckthorn which certainly grows at Newborough.

Bee orchid on dunes

Holiday caravans are one of the most profitable 'crops' that can be cultivated, and occur all along the Welsh coast, taking advantage of the attractive coastal locations and the access to wildlife, as well as the coastal resorts and beaches. Golf courses, as we have seen, are another very popular enterprise, using relatively large areas of land for leisure purposes. They can be profitable even in areas with low population, if there is a ready source of visitors. With sensitive and enlightened management, both caravan sites and golf courses can strike a good balance with the encouragement of wildlife. After all, the people who use these facilities are always glad to see the birds, butterflies, flowers and other wildlife of the coastal environment.

Early marsh orchids in the dune slacks

Excursion – Morfa Dyffryn

OS map Outdoor Leisure 18 – Harlech, Porthmadog & Bala / Y Bala
Total distance: 4 miles, mainly easy going and dry underfoot.

Park at Benar Beach by turning off the A496 about halfway between Barmouth and Harlech, just north of Talybont village. Signposted Traeth/ Beach, the road is called Ffordd y Benar. Continue to the end of the road and park in the car park. Walk westwards through the dunes on the marked path and wooden trackway, and turn right up the beach for half a mile or more.

This is an unofficial but well-recognised naturist beach, so may be busy in high summer!

Head north-westwards up the beach, following the edge of the dunes. You won't see any dune building here, as the sea is currently eroding the dunes. About 1 km up the beach there is access into the dunes on a right of way. If you visit in late spring/early summer you are in for a treat of wild flowers as you cross the dunes. You will see some of the striking plants that I have mentioned above and, in good years there are swarms of orchids, especially in the damper slacks. If you continue eastwards you will come, after about 600 m, to the edge of the Llanbedr airfield where some fairy ring mushrooms form enormous circles. A path follows the perimeter fence northwards, and crosses the dunes again onto Mochras (Shell Island). If you want to make a day of it, you can go all the way there through some swampy areas, and surprise the campers who thought they were on an island! However, you will have now seen a good cross section of the delights on offer here. There is a route back on a right of way that leads through the Dyffryn Seaside Estate, but the last time I went that way I had to climb the fence into the estate as the gate had been locked. It would be easier to return by the same route, enabling you to experience the marvellous dune environment for longer. This was a favourite haunt of the renowned botanist Mary Richards (1885–1977), whose life is described by William Condry in his book *Wildlife Safari – The Life of Mary Richards* (Gomer Press, 1998).

All dune systems have their unique features, and there are numerous locations all around the Welsh coast: here are a few example walks.

- **Point of Ayr** is the most northerly point of mainland Wales (Anglesey goes further north). You can walk there from Tan Lan (on the A548, 2½ miles north-west of Mostyn), along the sea wall for just over a mile, passing the old colliery, or you can park at Talacre (more caravan sites). The dunes start at Point of Ayr and continue westwards for several miles. If you go as far as Prestatyn you will come to the golf course after about 2½ miles.
- **Newborough Warren** has plenty of access from the village of Newborough in the south of Anglesey. In particular, you can park at a Forest Enterprise car park, 2 miles south-west

of the village (SH405634), and walk a mile along the beach to the delightful and romantic Llanddwyn Island, which is accessible at all but the highest tides. There are also dunes at **Aberffraw**, where there are important archaeological remains.

- **Ynyslas**, I have described briefly above. It can be visited by travelling north from Borth (passing the golf course). There is a car park and an informative Visitor Centre run by the Countryside Council.
- From **Pembroke** you can visit the fine and extensive dunes of the Angle peninsula at Freshwater West. Follow the B4320 westwards. After about 7 miles it crosses the landward side of some high level dunes, and a left turn on the B4319 brings you to the sea and a good car park with public conveniences. The Pembrokeshire Coast Path passes through the dunes here, at Broomhill Burrows.
- **Three Cliffs Bay** and Pennard Burrows are 4 miles west of Swansea University, on the B4436 from Black Pill to Pennard, where a minor road goes straight on past the church to Southgate. Park your vehicle overlooking the cliff and walk westwards on the cliff path. It is about a mile to the bay, which is surrounded by dunes. You may see horses galloping across the sands of the Pennard Pill estuary here. A footpath follows the stream, and splits into a network of tracks allowing access to the interesting ruins of Pennard Castle and church. A footpath through the golf course returns you to Southgate. This easy 2½ mile walk covers some of the most beautiful country in Wales.
- **Kenfig Burrows** can be experienced via a number of rights of way from the villages of Kenfig, 2½ miles north of Porthcawl, and Mawdlam, half a mile further north.

CHAPTER 4

Alluvial valleys & terraces

Picture a broad, meandering river lined with pollarded willows and punctuated with the occasional angler: a *Wind in the Willows* image of the lazy British river. Whilst England is well supplied with rivers of this nature, one has to seek them out more carefully in Wales. The mountainous nature of the country means that you are more likely to find a rushing stream flowing in a deep V-shaped valley than a broad gentle river. Such land *is* to be found, however, in most parts of Wales, on the Dee, the Clwyd and the Conwy in north Wales, for example; the Dyfi, the Wye and the Severn in mid Wales, and the Usk, the Neath and the Tywi in south Wales. The Clwyd and Tywi valleys in particular are famed in agricultural circles for their productive grassland soils. Parts of several smaller rivers also open out to give broad areas of alluvial flats as discussed earlier.

The larger river valleys have provided access routes to the remoter parts of the country for millennia. In south Wales the 'Valleys' is a term synonymous with industrialisation and urban development – indeed, many of these south Wales valleys are now floored not by alluvial deposits but by the spoil from coal mining (see chapter 13). These types of land differ from the estuarine areas described in the earlier chapter, although there is a transitional zone around the tidal limit. The Ordnance Survey maps will tell you where this point is: on 1:50,000 scale maps it is where the line demarcating the river turns from black to blue; on the more detailed 1:25,000 scale maps the place is marked by the abbreviation NTL meaning Normal Tidal Limit.

There is a variety of soils on river alluvium, most of which are stoneless. They range from very poorly drained to freely drained, and some of them, where protected from flooding, are among the best and most versatile soils in Wales. The Royal Welsh showground at Builth Wells, for example, is on beautiful deep, well-drained alluvial soils, called **Teme series**. They are brown and unmottled below the dark topsoil, and provide a good rooting medium that can be 150 cm deep. The silty texture of the material holds plenty of water for crop growth well into the summer, whilst its good structure allows sufficient excess water to drain through the profile, preventing waterlogging. A look at the National

Winter ash and oak on the lower Dyfi, near the tidal limit

Soil Map, however, will show that most river valleys are coloured blue – indicating wetter soils – mainly alluvial gley soils called **Conway series** (see photo). These have fine silty B horizons with mottled grey colours, rather similar in appearance to those of the Tanvats series on estuarine alluvium (see p.32).

Keeping your feet dry (see Box p.54)

We used to refer to gley soils as *poorly drained* a useful enough term, but one that has led to some misunderstanding. So, back in the 1970s the Soil Survey adopted a system called **Wetness Class** in which soils that are seldom, if ever waterlogged are placed in Wetness Class I and those that are almost permanently waterlogged in Wetness Class

VI. Classes II to V are intermediate. The classes are defined by the length of time soils are likely to be waterlogged at particular depths during a normal year. For example, Conway soils in agricultural use are usually in Wetness Class IV, a state that has been achieved by artificial drainage measures – without this they would be in Class V.

The Wetness Class of the soil has far-reaching implications for its use and management. What it means in the case of Conway soils is that the land readily becomes waterlogged after rainfall. During the late spring and summer excess water (from rainfall) in the soil is soon removed by evaporation from the surface, but more importantly by transpiration. This is the process whereby the roots of plants take in water and 'pump' it up through the plant to the leaves. It passes through the leaves as water vapour via tiny pores called stomata and disperses into the atmosphere. This has the effect of removing large quantities of water from the soil. As temperatures cool in the autumn, growth slows down at the same time as evaporation, and the process slows down to the point where rainfall inputs exceed the outgoings through transpiration and evaporation. At this point gley soils become waterlogged. Drainage through the profile can be improved by putting plastic or tile drains below the surface, and providing routes for the water to get into the drains. This is usually done by placing permeable gravels above the drains. In this way Wetness Class can be improved, usually by about one class. Knowledge of the Wetness Class of the soil in combination with regional climate can be used to predict at what stage the onset of waterlogging will occur in average years. This is the time at which the farmer should aim to remove stock from the land, as not doing so will result in permanent damage to the grass sward and, more importantly, to the soil structure. The action of hooves on waterlogged soil breaks down the peds, reduces infiltration and makes surface waterlogging even worse, so conditions spiral downwards leading to disaster. We have all seen fields where this has happened – they become a sea of mud. Once the surface structure has been destroyed in this way it will take many years to recover unless expensive amelioration methods are undertaken.

The phenomenon does not occur only in winter. On several occasions a similar thing has happened on the National Eisteddfod field when the event, as so often, has coincided with a period of wet weather. It has been necessary to provide wooden boards to enable the eisteddfodwyr to make their way across the Maes. Conditions have to be particularly bad for these problems to occur in summer – excessive treading, very wet weather and high Wetness Class soils. The organising committee have problems in finding a suitable site in many parts of the country, where they have to find a combination of a large area of relatively flat land with good road access and a willing landowner. Perhaps it is understandable that the

Wetness Class of the soils comes low down on the list of priorities.

To return, however, to the agricultural scene, the same considerations can be used to predict when, in the spring, it is usually safe to return stock to these wetter soils. The Soil Survey devised a scheme for calculating the best dates around which to plan the movements of stock to maintain sustainable use of soils for grazing. Indications are given in *Soils and their Use in Wales*. Most farmers have learnt over the years how best to manage their soils, but the figures are particularly useful to people buying new land or undertaking new enterprises. The Wetness Class of soils can be used in the same way to manage the timing of ploughing, or other mechanised activities on the land.

A place for nature

Whilst much of the drained alluvium is in Wetness Class IV, some of it is significantly wetter, particularly in old abandoned channels of the river. These may remain as small ponds and lakes known as oxbow lakes. Other areas, however have completely silted up and can be seen as slight depressions, whilst some can be identified only in air photographs. In any case, they are usually difficult to drain and form wetlands with rushes and other plants that provide wildlife havens for flowers, birds, mammals, insects and amphibians. Many of these areas now enjoy protection as nature reserves and

Rhiangoll valley

Alluvium and river terraces in Cwm Rheidol

Sites of Special Scientific Interest. Dolydd Hafren, for example, is a reserve of the Montgomery Wildlife Trust on the banks of the Severn near Montgomery (see p.58 for access details). It is an area of extensive wetland, including oxbow lakes. Hides have been constructed so that you can observe a variety of wildfowl and, if you are particularly lucky, otters. Some of the wetlands have been left alone long enough to form woodlands. Only willows, alders, and occasional birch can mature on the wettest of these sites, but sometimes there are drier islets and banks that will support other tree species.

Small differences in wetness class, and in the

degree of water movement through soil profiles will affect the wildlife of wetlands, whilst inputs of nitrates and other nutrients from surrounding fertilized fields also have an effect. Some unusual sites occur in river flood plains. For example, the stony alluvium in part of the Ystwyth valley has an unusual lichen flora on the shingle. There are other unusual areas where woodlands are frequently eroded by flash floods, creating distinctive dynamic woodland vegetation. Over the years gaps are created which then fill with new vegetation. The next flood may well erode another part of the land – a process called cyclic gap creation, something that must have been more frequent before the domestication of the countryside. Water meadows were fairly common in the past, but this form of management involving controlled annual flooding is now very rare.

As you might expect, some alluvial soils are intermediate in wetness between the poorly-drained Conway series and the well-drained Teme series. These are classed with the Brown Alluvial soils, but have greyish mottled horizons below 40 cm depth. They are called Gleyic Brown Alluvial soils.

Adjacent to the flood plain there are usually low terraces with more stony, freely-drained soils. These **river terraces** were formed at the end of the ice age, when melt-water from the glaciers was pouring through the valleys carrying a large burden of gravel. In the lower and middle parts of the valleys this gravel was deposited as the meltwater rivers lost some of their energy and spread out over the flood plain. The terraces are now above the level of the current flood plain, and have well-drained soils that are useful for arable agriculture. The most important of these soils is the **Rheidol** series. Although the parent material is gravel, the surface horizon is usually not too stony, and the particle size class is mainly clay loam, since mudstones dominated the gravels, and have weathered to give larger proportions of clay than are found in many gravelly soils elsewhere. Properties of these profiles are similar to those of the important Denbigh series, which is described in the next chapter. Many of the gravel terraces have been built upon, as they are good, flat, well-drained lowland sites, and the valleys represent useful communication corridors. The market town of Lampeter in Ceredigion is a typical example.

Gravels on river terraces can be of geomorphological importance, giving clues to processes that occurred during glacial and early post-glacial phases. Some are SSSIs on this basis – e.g. at Llan-non between Aberystwyth and Aberaeron where the gravels are being constantly exposed by coastal erosion.

Hartnup's laws of bogology

Walkers in Wales inevitably find themselves having to cross very wet places. Most frequently this is an ill-drained piece of track across moorland. The network of rights of way that crisscross the land has been derived from ancient rights of access and, more often than not, your path was once a thoroughfare of considerable importance to people for their day-to-day lives. Many of these old tracks once had fords crossing the minor streams that they intersect along their route. This is of more than academic interest because the fords were furnished – either naturally or by dint of hard work, with hard, stony beds so that the traveller or his/her horse did not sink into the mud. I have found a *rhyd caled* (hard ford) frequently in the middle of nowhere along the route of an unprepossessing right of way. Nowadays the stones are often hidden just below the surface of water or mud and wetland vegetation, but judicious poking with a stick will reveal them. It is therefore possible to cross these relatively dry shod – although you need to be careful that the stone that you choose to tread on is stable. In the old fords they usually are. Analysing my own actions and those of other walkers over the years, I have developed the following rules that may help when crossing boggy land. They assume that you are faced with a very wet area that must be crossed. Of course, it is sometimes possible to plan a drier route by scrutinising the layout of the land and selecting convex slopes, or areas of drier vegetation. For example, heather and the short, wiry mat grass (*Nardus*) are better than purple moor-grass (*Molinia*), bog moss and cotton grass or moss and rush flushes. Armed with the following information you have a substantially better chance of returning home with dry feet!

1 **Do not lunge!** This is the most important rule. People seem to think that if they rush across a wet patch they will spend less time in it and escape without wetting the feet. The opposite is true. It is better to apply weight gently – then if you seem to be sinking in, you can retreat and try another place. Tiptoe through the rushes! Speaking of which, it is often possible to fold over a tall clump of rushes and use them as a sort of 'snow-shoe' – to spread the weight of the foot.
2 **Poke with a stick** and seek out stones or firmer ground just below the surface – then tread exactly where you have probed.
3 **Study the vegetation**. The bright green and yellowish patches of bog moss (*Sphagnum*) are likely to be supported by nothing more than water or waterlogged peat. Darker patches of grass will be firmer. It is sometimes possible to tread on tussocks, particularly of deer grass and *Molinia*. One of the most difficult vegetation types to cross is the lines of rush and *Polytrichum* moss which follow permanent or temporary water courses. It is worth studying

these to identify exactly where the water runs – or you can find yourself in a very deep ditch. Sometimes you can select the route that sheep have taken through these.

4 **Trust your boots**. If you keep them well waterproofed you can stand on stones in streams that are submerged a couple of inches below the surface without getting your feet wet. It is better for you, your boots and the vegetation to do this than to try a diversion that goes through peat and/or acidic mud.

5 **Watch your companions**. See how they fare, then you can either choose to follow them, or select a better route.

6 **Wet feet are not the end of the world**. When I was a child I used to think that they were. The two worse crimes that I could commit in the eyes of my mother were coming home with wet shoes and socks, and coming home smelling of wood-smoke. (I am sure there *are* worse crimes, but I didn't commit them!) Washing and drying clothes and boots is much easier now than it was when I was a lad!

I don't want to end on a defeatist note, but sometimes it is just impossible to avoid getting wet feet. Western South Island, New Zealand makes Wales look arid. There they have many National Trails, and avoiding wet places by going around the sides of the trail is frowned upon because it causes erosion and gradually enlarges the wet part, so you are encouraged to splash straight through. The mountain huts fortunately have excellent drying facilities.

Llan-non – a mediaeval field pattern

The village of Llan-non is a particularly interesting example of a settlement built on Rheidol soils. According to G R J Jones (in *Mediaeval Villages* (1985) ed. Della Hooke), "Nowhere in Wales is it possible to gain a better impression of the form and pattern of Welsh mediaeval hamlet settlement than in the area known as Morfa Esgob in Llansanffraid parish" (i.e. Llan-non). The settlement now sits astride the A487 trunk road between Aberystwyth and Aberaeron, in the parish of Llansanffraid. It occupies an area of prized arable land between two streams draining the Mynydd Bach – Peris to the north and Cledan to the south. The soils (Rheidol series) are developed on loamy, well-drained gravels, and are the best soils for cereal growing in the moist climate of west Wales.

A glance at the O.S. 1:25,000 map shows the unusual field patterns of the area which constitute a unique historical document written upon the landscape. The area of long narrow fields represents the remains of shareland – or *rhandir*, a form of holding that goes back to the early mediaeval laws of Hywel Dda. The shareland in Llan-non is called *Rhandir yr Esgob* (Esgob = Bishop). The land was owned by the bishop of St Davids, but let to a grouping of relatives, which formed a *gwely* (or *lectus* in Latin), with distinct rights and duties under the Welsh laws. This *gwely* oversaw bondsmen who worked the land as serfs, but also had certain rights under Welsh law. Individuals worked their own disparate strips, or *lleiniau* (slangs in English), which have remained unfenced, as they were traditionally in arable use, separated only by low earth banks or balks. The complexities of mediaeval field systems in general are well described by Oliver Rackham in his classic work, *The History of the Countryside* (1986) – although all his examples are from England, where the laws were different. Many of the English systems have S-shaped curves that allowed the oxen plough team to begin its cumbersome turn before the end of the strip. The Llan-non strips do not all show this feature, but it is apparent in many of them.

The hamlet of Llan-non was held by the bishop under a traditional system in which rents, or 'tributes' were paid in kind as well as money. In 1326, for example, the occupants paid the bishop 6s. 8d. (33p) per annum, plus a cow every third year, and services of transportation and building work. It is the exceptionally traditional nature of the land holding under the bishops' jurisdiction, that has enabled the conservation of the ancient pattern of land holdings here. An ownership map shows how, even as late as 1841, a particular land owner, Evan Jones the mason, held 4½ acres, dispersed into 6 separate furlong-like strips. At that time, no less than 70 owner-occupiers were recorded for Rhandir yr Esgob. Even today the holdings are numerous and dispersed.

This historical digression shows how the nature of the soil in an area can, in exceptional circumstances, control its use and management for long periods of history. In the Netherlands, soil surveys were used as recently as the 1970s to work out the relative value of mediaeval strips, so that they could be consolidated, enabling farmers to swap strips with each other, and farm a single block of land.

Excursion – The Teifi at Lampeter

OS map Explorer 199 – Lampeter / Llanbedr Pont Steffan
Total distance: 3 to 4 miles, flat and easy going, but many stiles, and can be muddy.
Park in the Co-op car park at the southern edge of Lampeter on the east side of the main road, just north of the bridge.

Cross the road and turn right for a short distance, then turn left into a housing estate. When the road bends to the right, carry straight on down a lane that descends off the river terrace on which Lampeter is built, and onto the alluvial flat, where it becomes a footpath. You can now walk along the river and see all of the features that I have described above. You need to keep a sharp eye out for the stiles, as the river follows a very convoluted route here, and the footpath cuts off some large meanders. It is probably best to walk as far as you feel like, and then return by the same route, but you could go as far as the track to Dolaugwyrddon Isaf, and follow this northwards via Dolaugwyrddon Uchaf to the A475 at Pentre-bach, and return along the road. About half a mile into this walk, the river terrace has been eaten into by the river, and can be seen clearly on the south bank.

This walk is typical of many river walks, and alternatives can be found all over the country. Here are a few examples:

- **Afon Clwyd, Rhuddlan**: From the tidal limit at Rhydyddauddwr, where the Elwy and Clwyd join, you can follow the Clwyd south for about a mile and a half to Pont Dafydd beside the A55. There are also paths following the Clwyd northwards from **Ruthin**.
- **Afon Conwy, Trefriw**: From the centre of Trefriw paths lead north and north-west to the tidal part of the river. Once you have reached the river bank you can follow it south-eastwards, then cut the corner upstream of the tidal limit and follow the Conwy south – passing an excellent example of an oxbow lake – to a footbridge over the river to Llanrwst. You can return on a lane directly to Trefriw.
- **River Dee/Afon Dyfrdwy, south of Wrexham**: From Overton Bridge, 5 miles south of Wrexham on the A528, there are footpaths and lanes on both sides of the river going north. A good 5 mile circuit would be via Asney Park farm on the east bank, then climbing the steep wooded valley side at SJ380431, footpath south-west to Argoed, cross the road for footpath to Overton, then back on a footpath from the cemetery to follow the river westwards via Dee Bank, then north to the A539 and back to Overton Bridge.
- **Severn/Hafren at Caersws**: There is a short walk (¾ mile) along the Severn from the south side of the bridge at Caersws to the Maesmawr Hall Hotel. Here, as a bonus, you can see traces of a Roman road that led to the fort at Caersws.

- **Severn/Hafren at Montgomery – Dolydd Hafren**: Around the confluence of the Severn and the Camlad (SJ208004) is a nature reserve of the Montgomery Wildlife Trust called Dolydd Hafren (Severn Meadows). It is approached from a farm track at The Gaer (SO209997). This can be reached from Montgomery by taking the B4388 towards Welshpool for about 2½ miles, then turning left on a minor road through Forden village for 1¼ miles. The track, which can be muddy, heads north-west for nearly ½ a mile to the Reserve car park. Paths from here lead to two observation hides where you can watch typical wildlife of these alluvial areas. You may even see otters if you are both patient and lucky! The alluvial soils, away from abandoned river beds, are mainly the better drained Teme series, and, unusually, there is some arable cropping in this area.
- **River Lugg at Presteigne**: Right next to the town is a fascinating nature reserve belonging to the Radnorshire Wildlife Trust. Start at the car park and picnic area on the east side of the B4355, just south of the bridge, north-west of the town, or you can visit the area from the town using a riverside footpath. Here you will see some traditionally managed meadows on Conway soils, and an area of withy beds – willows, formerly coppiced and used for making baskets, hurdles etc. The wood has been allowed to grow into a willow wood and has a range of typical plants below the old trees. The place is a haven for all kinds of wildlife – you might be lucky enough to see kingfishers here.
- **The Wye at Glasbury**: Much of the Wye alluvium is well drained, and the valley west of Glasbury is no exception. A track leaves the B4350 just west of the village, and follows the riverside to Glasbury farm. It continues, across terraces and alluvium (Teme series) to Glangwye. You could return on a riverside path to regain the track just beyond Glasbury farm. This makes a circuit of about 4 miles.
- **The Tywi west of Carmarthen**: From Llandilo-yr-ynys, one mile south of Nantgaredig on the B4310 a right of way follows the north bank of the Tywi westwards for about a mile, cutting across the flood-plain northwards to the A40. From the same vicinity you can go north east from the inn, across the fields to Pont-ar-gothi, meeting the Afon Cothi which joins the Tywi just east of Llandilo-yr-ynys. You could make a 5½ mile circuit using these two paths, and linking them with lanes and paths to the north of the A40. Like much of the Tywi valley, this area is dominated by well drained Teme soils. The southern parts of Nant Garedig and Pont-ar-gothi are on Rheidol terrace soils.
- **Usk, Wye and Severn valleys**: The waymarked Usk Valley walk, Wye Valley walk and Severn Way all provide excellent opportunities for looking at the features of these rivers. Interesting sections are, from Llantrisant four miles northwards to Usk, and, on the Wye, the three miles south from Monmouth to Redbrook. Parts of Offa's Dyke path also follow the Wye, whilst the 10½ mile section from Buttington Bridge to Llanymynech is never far from the Severn or its tributary, the Vyrnwy.

CHAPTER 5

Lowland pastures and woods

The lowland areas of Wales are of an importance disproportionate to their extent. For a start, most of the population live in the lowlands, but perhaps more significant is the fact that all of the arable production and much of the stock raising takes place on lowland soils. There is a range of soils that often occur in complex patterns. Much of the lowland, especially in south Wales, is taken up with housing and industry. In parts of the valleys it is difficult to find natural soil unaffected by building or the wastes from coal mining. The countryside is generally predominantly rural however, and even close to built-up areas, the landscapes have a distinctive charm. The rolling slopes usually have a backdrop of hills, and the largely pastoral land has escaped much of the agricultural desertification that has homogenised large tracts of southern and eastern England. The settlement pattern, also, has led to a landscape of scattered farmsteads and intricate, confusing networks of ancient, often sunken, lanes. Our abundance of rainfall keeps the land green throughout most summers and the quality and clarity of the light is affected by variations in humidity which are a boon to artists and photographers.

School textbooks nearly always suggest that the main soil of Great Britain is something called a **Brown Earth**. In Wales these soils do not dominate the landscape because much of the country is mountainous, and Brown Earths occur only in lowlands. What is more, much of lowland Wales is covered with ill draining, heavy textured material that was dumped by the ice during the last glaciation, giving rise to **Gley soils**, as we will see below. Brown Earths are important though, because they are permeable and usually easy to work throughout the year, so they are valued for agriculture. They also support a much wider range of forest trees than can be found on wetter land. They are freely drained soils with well developed A and B horizons, often over bedrock of some kind. These types of soil are, not unreasonably, coloured brown on soil maps. They are fairly frequent in lowland parts of Wales unaffected by boulder clay (see below). The commonest types are the **Denbigh series**, developed over acidic sandstones and siltstones, and the **Milford series** over Old Red Sandstone. The Rheidol soils (discussed in the last chapter) are found on gravelly river terraces, and are Brown Earths too.

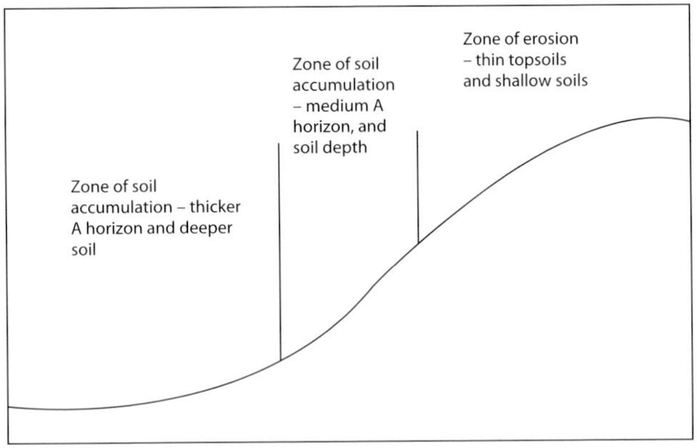

Typical slope sequence in lowland Wales (this is sometimes called a topo-sequence or catena)

Brown Earths have dark brown topsoils with loamy particle size classes and good structure, especially under grassland. Below the topsoil is a brown B horizon which lacks the grey colours and mottles so characteristic of gley soils (see below). The rich colour is the result of iron compounds, mainly complex oxides which, like rust, have a reddish-brown colour. Some soils are, in fact, red. The Milford series, which occurs on the Old Red Sandstone in Gwent, south Powys and Pembrokeshire, is red because the rocks from which it is formed are derived from strongly oxidised deposits that were laid down under desert conditions millions of years ago. The light textures and well-developed structure ensure that rainwater passes through the profile rapidly enough to prevent serious waterlogging.

Most of the Brown Earths in Wales are acidic (see Box p.64). The pH in the topsoil tends to be higher than in the subsoil as a result of the addition of lime over the years. The Welsh coastline is dotted with harbours, large and small, used and disused, and most of them have a limekiln or two, a reminder of the time when it was vital to 'sweeten' the land with additions of calcium carbonate. Limestone from Pembrokeshire was transported along the west coast as far as the Llŷn peninsula where, in the late nineteenth century, a road, 'Y Lôn Goed', was especially constructed by one landowner to transport the lime inland. In general, the wetter the climate, the more acidic the soils. This is because rain tends to wash the alkaline bases out of the soil. Of course, the parent material also has an effect, and the hard acidic rocks of mid and west Wales give rise to more acidic soils than do the softer Old Red Sandstones. The younger rocks of the Vale of Glamorgan are mainly limestones, and their soils will be discussed at the end of this chapter.

The landscapes where these lowland soils occur are typically undulating, and interesting variations in the profiles relate to the slopes where they are found. We think, perhaps, of soils as static and unchanging, but in fact they are never stationary. The processes of weathering and plant growth that were responsible for the formation of soils from bare parent materials in the first place are still going on. This is most easily seen on a hill slope. The top of the hill is usually convex, and it is here that most erosion is taking place – upper slopes and summits are more exposed to wind and rain, and gravity is slowly but surely moving the topsoil down the

Cultivation terrace near Rowen

hill. Thus, soils on the brow of the hill tend to be shallower than those in mid-slope positions, where soil is moving down, but being replaced by material from above. At the base of the slope we usually find a concave area where the eroded soil has accumulated. Here the topsoils will be significantly thicker than elsewhere.

If there are obstructions like walls and hedges across the contours, the soil will be banked up against the up-slope side, and may be eroded away from the down-slope side. Prehistoric farmers made terraces on steep slopes by placing boulders along the contours and filling the gaps with smaller stones. Ploughing of the land between these proto-walls accelerated the down-slope movement of soil, banking it up against the obstruction, which could then be built upon until a relatively level terrace had been created. Cultivation terraces of this kind can still be discerned in some parts of the country, particularly along the western coasts, for example on the well-drained soils above Dyffryn Ardudwy and Llanbedr in western Meirionnydd.

Early purple orchid with wood sorrel and bluebells, Pembrokeshire

The well-drained, relatively fertile Brown earths are mainly in fairly intensive agricultural use, but some areas have small woodlands, with oak, ash, sycamore and occasionally, where the land has been in woodland for several hundred years, small-leaved lime, with potential for a rich ground flora. The more acidic woodlands have bluebells – a plant that is so familiar to us that we forget that our bluebell woods are one of the glories of Europe, and confined to its western, Atlantic fringes. The less acidic areas have ramsons (wild garlic) and wood anemone, whilst the least acidic may have dog's mercury. Early purple orchid is a spectacular addition in some of the older woods on Brown Earths (these are particularly abundant in the lime woods along the Wye valley, and can be seen, in late spring, along the Offa's Dyke National Trail).

Some of the pastures and hay fields are full of colourful herbs, although many of these traditional fields were ploughed and reseeded with ley grasslands following the Second World War. The best hay field floras are now found in cemeteries, where the cutting of the grass resembles traditional hay management. Currently, attempts are being made to re-create species-rich grasslands. The Gwent Wildlife Trust, for example has acquired some 100 acres of agricultural ley grassland on Milford soils (see above) on the plateau above the right bank of the Wye, south of Monmouth. They are embarking on a 100 year programme to reduce fertility in these fields and develop a mosaic of wood pastures and species-rich hay meadows, using their reserve at Pentwyn Farm, which has preserved glorious flowery hay meadows, as a template, and a source of seeds. Close attention to soil factors will improve the chances of success in this difficult venture. The presence of shallow, stony, droughty patches within Brown Earth map units can increase biodiversity, especially where these areas have remained unploughed. I have more to say about hay meadows in chapter 8.

Gley soils

A glance at the soil map shows that, even in the lowlands, Brown Earths are by no means the commonest soils in Britain, and many areas of the soil map are dominated by green colours representing Gley soils. This is the soil classification term for soils wetter than Brown Earths, that tend to have excess water in the topsoil for long periods of the year, as we discussed when talking about estuaries earlier. The determined country walker will not be surprised to hear that these wetter soils are common: any ramble, even in the lowlands, seems to leave one with decidedly muddy boots. Welsh mud, however, I have discovered to be significantly less sticky than English mud! Walks across ploughed fields in the Cotswolds, Chilterns and many parts of Kent, for example, involve a rapid build-up of sticky deposits on the boots, leaving your feet twice as heavy as they ought to be. In northern England the situation seems to be more like that in Wales – the boots get very dirty and muddy, but the stuff does not stick like glue. On the whole we have the ice ages of the past two million years to thank for our wet walks. They left great swathes of the country covered with a ground-down heavy textured material called boulder clay – more properly known by its (less descriptive) Scandinavian name – ***till***. This material tends to cover the rocks in hollows and valleys, and on many plateaux. It is dense and slow to drain, so the soils developed in it are **Gley soils** (see p.33). Like the alluvial gleys described in the chapters on estuaries and river valleys, they have B horizons in which most of the iron content is in the reduced (ferrous) form, giving an overall grey colour to the body of the soil. Re-oxidation in small parts of the soil aerated by cracks and roots produces flecks or **mottles** of rusty red, orange and yellow colours. They are called Surface-water Gley soils, since the wetness is caused by water which is held in the upper horizons of the soil and is unable to pass quickly enough through the slowly permeable subsoil. Unlike alluvial soils, they are usually fairly stony – as the name boulder clay suggests.

Soil acidity

Gardeners and farmers often talk about acidic and basic, or alkaline soils. How many of us actually know what these terms mean? We know that hydrochloric acid is nasty stuff that will burn the skin, and that vinegar and lemon juice are weak acids that are piquant, but will not harm the mouth, whilst bleach is dangerously alkaline. It is the highly active hydrogen ions, types of free radical, that cause problems like the burning property of acids. Pure water with nothing dissolved in it has a very low proportion of these hydrogen ions because very few of the H-OH molecules dissociate to release H^+ and OH^- ions. Water is a very stable compound in which the atoms are strongly bound to each other. The ions are at a concentration of one to every 10 million water molecules. This is 1/10,000,000 notated mathematically as 10^{-7}. Acids are less stable compounds that dissociate and release hydrogen ions into solution. The greater the concentration of these ions, the more acidic is the solution, and the more reactive.

By convention, acidity is measured on the **pH scale** – from the French **p**uissance d'**H**ydrogen – strength of hydrogen. It is inconvenient and cumbersome to speak of concentrations of 10^{-7}, so neutral water is considered as pH 7. As the scale is derived from a negative, the lower the number, the more acidic the substance. Hence a solution with 1/100 concentration of H^+ ions has a pH of 2 (10^{-2}). Concentrated hydrochloric acid has a pH of 1, whilst weak acids like vinegar and lemon juice are pH 2 to 3. It is worth remembering that each point on the pH scale represents a tenfold difference – in this example the hydrochloric acid is ten times more acidic than the lemon juice.

The acidity of soils is very variable (in Wales, generally from pH 3.5 to 8). It depends on a great many factors, not least the nature of the parent material. Limestone (and chalk, a type of limestone which is absent in Wales) are very rich in carbonates that mop up excess hydrogen ions – so soils on these materials tend to be alkaline – pH 7 to 8. The presence of carbonates, especially calcium carbonate, in the soil can be detected by testing with a 10% hydrochloric acid solution – it will fizz and bubble, releasing carbon dioxide if it reacts with free carbonates in the soil. Free carbonates originate from broken down limestones, or from shell fragments – like those in blown sand – see p.38. There is also a relationship with climate: soils in wet climates are leached. That is to say the carbonates and other soluble minerals in the upper horizons have been washed lower down by the constant percolation of slightly acidic rainwater through the soil. The presence of large proportions of organic matter, humus and peat, will also lower the pH, as these substances contain natural organic acids. Finally, the vegetation growing on the soil itself affects the pH. Many heath species, like heather, release acidic substances from their roots, acidifying their own environment, and making it difficult for other species to compete. Coniferous foliage tends to filter acidic pollution out of the air which is washed into soil and streams when it rains, lowering the pH.

There are a number of reasons why pH is important in soils. The reactivity of hydrogen ions in the soil solution affects many of the chemical reactions that are going on, including the ways in which major plant nutrients like nitrogen, phosphorus and potassium are released from the clay minerals and humus to which they bind. It also influences the solubility of some of the minerals, which can be absorbed by plants only in the soluble form. Many metallic minerals like aluminium and manganese become more soluble at lower pH, to the extent that they can reach toxic levels. Plants that are to grow in low pH soils must evolve strategies for dealing with these toxicities and with the other conditions brought about by acidity. They are so sensitive to pH that it is frequently possible to estimate the acidity of the soil from the vegetation growing upon it. Sometimes, where there are pockets of deeper and shallower soils over calcareous rocks, you will find leached soils next to calcareous ones, and acid loving plants can be seen growing next to lime lovers.

Farmers from very early times became aware that adding carbonates – lime – to certain soils would sweeten the land, and make it more productive. We now know that most crops, including grass, grow best at pH 6 to 7.

Acidity affects the fauna as well as the flora. Acid soils lack calcium carbonate because acids react with it to form other calcium salts and carbon dioxide. The all-important earthworm is reliant on a certain amount of carbonates, and where there are lots of earthworms you will find their predators like many bird species, moles and badgers. Snails also rely on the presence of carbonates in soil and plants. Like the marine organisms that originally helped form much of the world's limestones, snails *scavenge*, or concentrate calcium carbonate from the environment to make their shells. The snails in my own garden, on the acid soils near Aberystwyth, have remarkably weak shells – because there is not much lime in the soil. Despite this, however, the saxifrages that grow in the garden have excess carbonates, which they excrete to form minute chalky blobs on the edges of the leaves. Even the rare and beautiful purple saxifrage, which grows on the slightly less acidic volcanic rocks on parts of Cadair Idris, does this trick, which must have an advantage for the plant: presumably it makes the leaves less palatable to grazing animals.

In Wales much of the lowland, where rainfall is relatively low (less than about 1,400 mm), is occupied by Surface-water Gley soils of this kind. There is a large variety of soil series depending on the parent material, the degree of wetness, the texture, stoniness, depth etc. The most common series are called **Cegin** – which is a silty clay loam, and **Brickfield** which is a clay loam in the same kind of material. Their profiles cannot be told apart in a photograph, as they differ only in texture – the distinction is largely academic anyway since they behave in a very similar manner when cultivated or grazed. Where the till is affected by the Old Red Sandstone, reddish soils called **Fforest** series are found – east of Ammanford for example. All of these gley soils occupy undulating, rather irregular ground, but slopes are usually fairly gentle. In some districts they are on till plains that cover considerable areas – east of Montgomery is an example – whilst the land around Llandrindod Wells is similar, with some particularly heavy gley soils, dominated by the clayey **Hallsworth** series east of Llandegley rocks, and along the Edw valley. In most parts of Wales, however, the till covers the gentle slopes of valley sides, and merges into brown soils as you climb out of the valleys.

Surface-water Gley soil from near Llandrindod Wells

In both lowland and upland landscapes, direct rainfall is not the only source of soil wetness. Springs and seepages are common, with spring lines often following the contours near the junction of the Surface-water Gley soils and the Brown Earths. Around the springs are groundwater gley soils affected by waterlogging from below the soil surface.

In early prehistoric times the lowlands would have been covered with forests. The brown soils would have been dominated by oak, with a rich mixture of other species including birch, lime, hazel, rowan and, especially in south Wales, elm. There may have been some ash on damper, but reasonably fertile soils. On the gley soils there would have been some oak and ash, but considerable amounts of alder, with willows in the wettest places. The sycamore, which also thrives on wetter soils, was introduced to Britain in Roman or mediaeval times.

By the Neolithic, about 5,500 years ago, the clearance of woodland for agriculture was well underway, and the process has continued ever since. The practice was of great importance to mediaeval people, as populations increased, and agriculture developed. The *Mabinogion* is probably part of an early medieval oral tradition that was written down in Welsh in the high mediaeval period, and appears in early fourteenth-century texts. In a related story, *Culhwch and Olwen*, the giant Ysbaddaden Bencawr sets the hero Culhwch a number of increasingly impossible* tasks including the following which was to be accomplished in one day:

> *A weli di'r llwyn fan draw? … Fe fynnaf ei ddiwreiddio o'r ddaear a'i losgi ar wyneb y tir hyd oni fyddo golosg hwnnw a'i ludw yn dail iddo, a'i aredig a'i hau nes y byddo'n aeddfed yn y bore erbyn amser codiad gwlit…*
>
> *Do you see the large thicket over there? … I want it uprooted from the earth and burned along the ground so that the cinders and ashes fertilise it; and I want it ploughed and sown so that it's ripe in the morning by the time the dew disappears…*

(Translation by Sioned Davies, 2007)

The chores of scrub clearance and reseeding have been familiar to farmers from Neolithic times to the present day, and it is interesting that by the time the material was recorded the idea of fertilising the ground with ash was well established. The giant's remarkably concisely phrased assignment also introduces the modern concepts of soil available water and moisture deficit. A newly planted crop will not thrive without a good supply of water for the roots. Deep, loamy, well-structured soils will provide this, but only if the moisture supply is replenished at regular intervals. The giant realises that once the dew has dried on the surface, and the sun begins to suck moisture from plants and soil, only a well-developed crop, with a good, extensive root system will survive.

Clearance of woodlands accelerated with the development of better tools during the Bronze and Iron Ages, and the expansion of the population. The situation today is that about 14 per cent of Wales is wooded. Most of this is coniferous plantations – semi-natural and planted broadleaf woodland covers only 1.3 per cent of the land.

* Needless to say, despite being 'impossible' Culhwch and his friends manage to accomplish the tasks with a heady mixture of magic, panache, imagination and luck, but that's another story…

The Vale of Glamorgan – decidedly different

The Vale is the most southerly part of Wales and with the coalfield, the most densely populated. Its soft rolling countryside, fertile soils and mild climate have attracted settlers for millennia. The patterns of villages and farms are more like the English lowlands than any other part of Wales. It differs geologically too, being underlain by a series of limestones and shales of Lower Lias age. These are Jurassic rocks, about 200 million years old, but nonetheless much younger than the rest of the country. Between these rocks and the coalfield to the north is a broad band of Carboniferous Limestone. The soils on both of these are mainly leached; that is to say the calcium carbonates from the parent materials have been washed out by rainfall over the millennia. However, the material still has a good supply of natural nutrients, as well as responding well to fertilizers as a result of its deep, loamy character. The Vale should more properly be called a plateau, as it is gently undulating. It is a pleasant, rural area of large fields and mixed farming. There are few woodlands, with Hensol Forest, east of Welsh St Donats, the largest. Typically, the forest cover survives on the more poorly drained soils, where agriculture is more difficult and less productive.

The soil pattern is quite intricate, but essentially there are two main soil series in the Vale of Glamorgan. The extensive Lias limestones and shales of the southern section are dominated by the **Ston Easton** series, a well-drained loamy, rather stony soil which becomes clayey below 40 or 50 cm, and passes into limestone between 60 and 80 cm down. The remnants of calcium carbonate in the subsoil ensures a good structure and permeability despite the heavy texture. Studies of the minerals in these soils have shown that, although they are on limestones and shales, there is a significant contribution of material dumped by the Irish Sea glacier, and also a little sand derived from the coastal dunes.

The Carboniferous Limestone is distributed mainly in the northern part of the Vale, although a strip extends eastwards from Cowbridge through St Lythans to St Andrews Major. These areas have lighter soils, called **Malham** series, which are moderately deep, silty clay loams over hard limestone. A few shallower, often very stony soils occur in this area on convex slopes and ridge tops. Ston Easton and Malham soils are both of good agricultural quality, and useful for both arable and pasture.

As well as these extensive soils, there are smaller areas of the Vale with soil types of particular interest. Around Hensol forest for example, there is a plain formed from deposits laid down by glaciers, and by the water draining from them. This has led to a hummocky terrain, sometimes called kettle moraine, as there are rounded depressions called kettle holes, which were left when, during the melting of the glaciers some 12,000 years ago, detached lumps of glacier ice were incorporated into till and glacio-fluvial deposits and eventually melted.

Hummocky kettle moraine near Welsh St Donats

Incidentally, it seems to me that the depressions are more like bowls than kettles, but perhaps 19th century kettles looked like bowls. The resulting landscape is a complex of gravelly ridges and wet hollows. The ridges usually have well-drained rather stony soils that we have already met on river terraces, called **Rheidol** series (see p.53). In the hollows the groundwater affects the soil profiles to a varying extent, giving wetter soils. Although the sandy loams of these soils are quite permeable, they are difficult to drain, as improvements would have to rely on lowering the regional water table – not something that can easily be done. And even if it were done, it would cause problems of droughtiness for the better drained sites. The complexity of this area is such that the soil series can't be mapped separately, so, the solution is to map what is called a *soil complex*, in which the constituent soils are

A woodland path in the Wye valley

described in relation to topographical aspects of the map unit – as I have done above.

Another interesting area is around Dinas Powis, west of Penarth. Here there are soils found nowhere else in Wales. They are called **Pelosols**, and have formed in Permo-triassic mudstones which weather to give a very clay-rich reddish material. What is unusual is that although very clayey, the soils, known as **Worcester** series (guess where else you can find them?) are moderately well drained. They are dominantly on convex sites, and the nature of the clay, combined with a relatively dry climate, creates fine cracks and a good structure, allowing infiltration and adequate drainage through reasonably permeable subsoils.

There is a fairly detailed soil map of the Vale, and an associated book by C B Crampton, *Soils of the Vale of Glamorgan*, published in 1972, before the Soil Survey had rationalised its classification. If you study this publication you will find the Ston Easton and Worcester series described, but the Malham series was called **Lulsgate** at the time.

Walks in lowland fields and woods

There are so many footpaths from farm to farm, village to village, and through the woodlands of Wales that I have not given any specific walks in this section. Every community has favoured walks along old tracks and field paths. A look at the map of your area will reveal these rights of way. Not so long ago many paths in agricultural areas were in a poor condition, and often impassable, but the County Councils, the National Parks and the Ramblers' Association have worked really hard over the past years to improve footpaths and stiles, while schemes like Tir Gofal have made large areas of farmland accessible to the public. Grant monies have been made available in recent years to improve these tracks and to put up interpretative panels on farmland, wild habitats and woodland, whilst the Woodland Trust is worthy of especial mention for its hard work and public spirit.

So when you walk these paths think about the kind of soils that support your feet, the effects that they have on land use and vegetation, and how the earliest prehistoric settlers would have viewed and used the resources supplied by the landscape. There is often a pleasing continuity revealed by the ways in which modern communities are interpreting and opening up the landscape for us to enjoy.

CHAPTER 6

Ceunentydd – the deep gorges

You are never far from a gorge in Wales. We may not have a Grand Canyon or even its European equivalents, like the magnificent Gorges de Verdon or Gorges du Tarn in France; but as the picture opposite shows, there are some pretty spectacular clefts through the rocks of Wales. Some of my favourite walks are in these deep, wooded, mysterious *ceunentydd* as they are called in Welsh (singular is *ceunant*). They are good places to go on windy, rainy days, as they offer some shelter, and the rivers are swollen and impressive. Several years ago I was on my way home from a surveying trip up north, and took the opportunity to examine soils around the Bala fault, in an area that I needed to check for the National Soil Map. I tried to cross a ceunant stream near Rhydymain, and was not careful enough. I slipped on a mossy rock in the stream bed and tumbled inelegantly into the deep, clear water. It was shockingly cold, but I remember that it tasted sweet and good. Fortunately the sun was shining and the river was relatively gentle, so I was able to scramble out unharmed, and drip my way back to the car, which was parked in a recently cut hay field. I was able to strip off behind some stacked hay bales and change into the dirty clothes that I had been wearing the day before! But I still like gorges.

I'm not certain about the etymology of the word *ceunant*, but suspect that the Welsh term comes from a combination of *cau* (closed) and *nant* (stream); and is hence descriptive of these evocative, distinctive habitats which give a precious feeling of remoteness even when close to roads and habitations. Most of our numerous gorges are the result of streams and rivers deepening their channels by processes of river erosion. It takes enormous energy for water to erode and cut through the landscape. This energy depends upon the volume of water, and the height from which it falls. In most of our rivers both of these increased dramatically during and immediately after the ice ages. Around 20,000 years ago, vast amounts of ice were locked up in glaciers, which covered all the higher parts of Wales, and formed a huge ice sheet in the area now occupied by the Irish Sea. As a result, the sea level fell by as much as 100 m and Britain was joined with Ireland. As the glaciers melted, about 12,000 years ago, they released huge quantities of water that flowed in old channels, and in new courses determined by the topography of the icescape. At the same time, the

Upper Ceunant Cynfal, near Ffestiniog

actual land surface, which had been compressed by the weight of the ice, rebounded and gradually rose (it's still doing so in Snowdonia). The combination of a lower sea level, a rising land surface and huge amounts of water allowed the rivers to erode deep gorges, cutting back from the sea. Geologists call this process ***rejuvenation***.

Steep, narrow gorges extend along the middle and lower parts of the courses of many Welsh rivers and streams. For example, a number of streams drain the uplands between Porthmadog and St Davids and flow into Cardigan Bay or the estuaries. They tend to be deeply cut or *incised* into the landscape as a result of rejuvenation. Rushing streams pour through the valleys over beds of rock, often forming cataracts and falls. The rocky stream beds and their surrounds are shady and mysterious, the boulders often concealed under a variety of mosses and lichens, and the gnarled boughs are festooned with ferns. The steep slopes usually remain thickly wooded, since they have always been too difficult to cultivate. Good examples are numerous, from the Llennyrch draining Llyn Trawsfynydd to Cwm Gwaun between Newport (Pembrokeshire) and Fishguard. Many of these woods have been exploited for charcoal manufacture since Roman times. (See William Linnard's *Welsh Woods and Forests* (2000), National Museum of Wales.) Trees include ash, oak, birch, hazel, hawthorn and the rarer small-leaved lime, a tree that reproduces by suckering, and can be very old indeed. It would not seem unreasonable to include these types of woodland as remnants of temperate rainforest, a vegetation type generally reserved for areas with more than 2,000 mm of rainfall. Although few of our ceunentydd are subjected to this much rain, the wetness is enhanced by the generally moist conditions and shady, sheltered sites.

The slopes of the ceunentydd in acidic rocks are dominated by Brown podzolic soils and shallow soils over rock or scree, known as **rankers**. The trees insinuate their roots between cracks in the bedrock, or through the bouldery screes. **Brown podzolic soils** are freely drained and resemble Brown Earths except in one respect – the subsoil is a bright orange-brown colour. The colour is so distinctive that in some parts of Britain it is called *fox-earth*. This is because, in upland areas, the subsoil is often exposed at the sides of roads where rabbits or other mammals have been digging into the bank. Perhaps it is also a reference to the colour of the fox. The reason for the bright colour is the amount of iron in the soil. We will meet these extensive soils again when we discuss the *ffridd* lands in the next chapter, where I will explain why they are found on steep slopes in the oceanic regions.

Brown podzolic soils are ideal for the sessile oak (*Quercus petraea*) which is frequently partnered by birch. The diversity of the gorges is increased by flushed areas of groundwater gley soils (see p.33), and sometimes boggy patches with humus-rich soils and peats. These are especially common in gorges that have been partially infilled with gravelly deposits (laid down by torrents of water

The deeply incised valley of the Ffynnon-ddewi at Cwmtydu

from the melting ice of the last glaciation), giving a flat area in the bottom of the valley. In the humid woodlands most of the soils have fairly large amounts of organic matter in the surface layers.

Paths through the gorges (where they exist) never seem to travel in a straight line. This is because below the trees the ground is broken up by outcrops and boulders, which are usually covered with bilberry and heather, common cow wheat with its yellow flowers, and the fine-leaved wavy hair grass (*Deschampsia flexuosa*). Deeper soils have scattered bracken, hawthorn and holly. There is a huge variety of mosses on the woodland floor and on boulders and tree trunks. Unfortunately, the mosses do not have consistently used common names (although the British Bryological Society has recently published a list of recommendations), so we have been using the somewhat daunting, but often elegantly beautiful Latin names. Four of the most characteristic are the pale hummocks of *Leucobryum glaucum* (smaller white moss), the attractive and bristly looking *Rhytidiadelphus loreus* (shaggy moss), the tamarisk-like branching *Thuidium tamariscinum* (common tamarisk moss), and the stoutish, shaving-brush fronds of *Dicranum majus* (greater fork moss). Boulders carry complex and attractive communities of mosses, lichens and liverworts often dominated by the close-growing moss, *Brachythecium rutabulum*. They look like artistically impressionistic tapestries in shades of green, yellow and brown, with dark red or purple strands provided by delicate leafy liverworts. Polypody fern grows on the boulders, and also as an epiphyte on the branches of the trees. The frequency of epiphytes, especially lichens, is an indicator of the clean air on our western coasts. Other ferns are common, especially the notable, shiny, dark green hart's tongue, which can be very prolific, and the hard fern with its distinctively different fertile fronds.

The wet valley bottoms may have purple moor grass (*Molinia*) bogs, or areas of swampy willow and alder woodland with yellow flag and kingcup giving welcome, and sometimes startling splashes of colour. The seepage areas, known as flushes, are often showily obvious as lines or patches of golden saxifrage, its foliage a uniquely bright greenish yellow, less exotic than the flags and kingcups. Elsewhere, large tussocky sedges make the going very difficult in these often remote and lonely places. They may also be dangerously boggy, or, where subject to nutrient-rich run-off, infested with nettles. Take heart that the difficulties of access for humans are a boon for the wildlife! The boggy vegetation of the flushes and valley bottoms renders an extremely useful service to the environment. It has been found that acidic water running into the flushed areas is modified by the root zone of the vegetation, especially the tightly packed roots of purple moor grass, so that when the water emerges to enter the stream or river, it is significantly less acid. Acid water entering streams can kill fish, and the insects that they eat; so we need to respect these types of vegetation, and avoid disturbing them.

Wales has many fine examples of these semi-natural gorge woodlands that are characteristic of the western fringes of Europe. In this oceanic zone, temperatures are fairly moderate throughout the year due to the influence of the Atlantic which acts as a kind of storage radiator, absorbing heat in the summer and re-radiating it during winter. Similar environments can be found all along the western seaboard of Europe. In moist parts of northern Spain, like the northern slopes of the Cantabrian Mountains for example, the latitude makes for a

Cwm Einion, Ceredigion

warmer, but familiarly wet climate, and similar sessile oak woodlands occur at altitudes of about 600 m (2,000 ft), where temperatures are similar to those in Wales.

Another feature of the oceanic climate is the more or less constant humidity. Shady, perpetually moist habitats favour a specialist flora for which west Wales is famed. As well as the shade cast by the trees and the confined nature of the narrow gorges, humidity is often increased by the splashing of waterfalls and rapids, and by seepages from the rock. Such conditions are ideal for mosses and liverworts. It is notable that although Britain has only about 12 per cent of the flowering plants present in Europe as a whole, a massive 60+ per cent of European mosses are represented. Similar woodlands are found on some upland crags, especially those that face north or north-east, where the shady, moist conditions of the ceunentydd are replicated. The gorge woodlands provide important breeding sites for the red kite, and, if you are quiet and patient, you might also see buzzards, redstarts, pied flycatchers and wood warblers.

Wild rhododendron is a beautiful plant with an exotic bloom, but it is a killer. It readily spreads from parkland and gardens, and once it gets a hold in our oak woodlands, it smothers the native ground flora. As well as suckering prolifically and rapidly covering the ground beneath the trees, it exudes chemicals that are able to inhibit the growth of competing plants. It seems to like well-drained, even droughty land just as much as wet bogs. In many areas conservation volunteers are doing sterling work uprooting this invasive plant and giving the rich diversity of native plants a chance to re-colonise.

The ceunentydd have a powerful effect on the imagination, and they seem to have inspired several folk tales. It was in the gorge of the Dugoed, a few miles east of Mallwyd that the infamous Red Brigands ambushed and murdered the local JP in the eighteenth century. As a result they were rounded up and hanged. In Ceunant Cynfal, south of Ffestiniog, tales from the *Mabinogion* tell how the dastardly Gronw Pebyr tried to kill Lleu Llaw Gyffes. He failed, and Lleu was able to exact revenge by shooting an arrow at him which passed straight through a boulder – the perforated boulder is said to have been visible on the banks of the Cynfal, although I have never been able to find it.

Excursion – Cwm Cletwr

OS map Outdoor Leisure 23 – Cadair Idris & Bala Lake / Llyn Tegid (SH659922)
Total distance – a little less than 1.5 miles; mostly flat and easy.

This short trip will allow you to see a typical ceunant, but one that has an unusually rich and rewarding heritage of wildlife. It is best seen in the spring, when the river is full, and the light filters through the fresh new leaves. Some sections of soil are exposed on the cliff walls of the valley.

The long, narrow wood in Cwm Cletwr at Tre'r-ddôl (Ceredigion) is a Site of Special Scientific Interest, and part of the northern side is a nature reserve of the Wildlife Trust of South and West Wales. It has a richer flora than most woods of the neighbourhood, partly due to its past management. Diversity is also enhanced by the variety of habitats, including very moist shady areas near the stream, steep banks, and seepages of water which are often enriched with nutrients from the rocks of this area. The valley bottoms are typically shady and moist, the damp climate being locally exacerbated by splashing from rapids and waterfalls, and by upwelling groundwater, which augments the main stream. These seepage zones are usually called *flushes*, and are often marked by carpets of golden saxifrage.

Although sessile oak is dominant over most of the valley, there is a good variety of other trees including the stately small-leaved lime, spindle, and some areas of ash wood with a rich ground flora including sanicle. The increased base status of these soils has been tested by pH readings, but also is evident from the presence of plants like dog's mercury, yellow archangel and sweet woodruff. As well as having a total of 140 flowering plants recorded, the area is particularly good for ferns (including hay-scented buckler fern, beech fern and filmy ferns), mosses, liverworts and lichens. The particularly moist microclimates of these valleys enhance the processes already described to give some very special niches.

Like most semi-natural woodlands, Cwm Cletwr gives excellent habitats for birds. Buzzard, sparrowhawk, pied flycatcher, redstart, nuthatch, wood warbler, dipper, grey wagtail and three species of woodpecker are known to breed, and I have seen both long-tailed tits and goldcrests in the same birch trees in the nearby woods of Cwm Llyfnant. It is possible that dormice feed in the wood, and there are brown trout in the stream.

Ceunant walks in other parts of Wales

- **Afon Meirchion, Henllan, Denbigh**: From the church at Henllan, two miles north-west of Denbigh, descend north-west on a minor road to the Meirchion and you can follow the stream through the woods northwards for just over a mile to its junction with the Elwy. You can return on the same path, or follow a lane north-west to Bontnewydd (where there is a pub) and visit the site of the famous Pontnewydd cave which revealed some of the oldest artefacts found in Wales. Protected in the cave from the Ice Ages were tools and human remains from about 222,000 years ago. Unfortunately this important site is not accessible to the public, so you will have to view it from the road. The exposed rocks on this walk are limestone, so have a look at chapter 10 before you go!

- **Afon Llugwy, Betws-y-Coed**: From the centre of Betws-y-Coed a path leads west along the north bank of the river. Within about a mile you come to the interesting Miners' bridge. The deeper parts of the gorge begin west of here, so you can walk as far as you like, and return by the same route, or devise a circular route using the numerous well-marked trails of the area.
- **Ceunant Cynfal**: A delightful circular walk of about 3½ miles can be made from Pont Tal-y-bont, between Llan Ffestiniog and Maentwrog. Go up the northern side of the stream towards Llan Ffestiniog, but once out of the woodland, turn to the right and descend towards the gorge. The footpath follows the stream to Rhaeadr Cynfal, where a rustic bridge gets you to the south bank, and you can return on a good path (muddy in places) to the start.
- **Tal-y-Bont** (Meirionnydd): From the car park in the village follow the northern bank of the river Ysgethin for a mile and a half to Pont Fadog, where there used to be a drovers' inn at Llety Lloegr. Before crossing the bridge you could visit the burial chamber, 350 yards to the north-west, then return on field paths and tracks to Tyddyn-y-felin (SH596218), and down the winding lane back to the start.
- **Cwm Gwaun, Pembrokeshire**: The river Gwaun is a deeply-incised glacial meltwater channel that runs into the lower town at Fishguard. There is a good walk of a mile or so up the north side of the river from Llanychaer, two miles south-east of Fishguard. You can cross Cilrhedyn bridge and return on the south side on lanes, or extend the gorge experience by continuing another mile and a half along the riverside lane to Pontfaen, where there is a pub.

CHAPTER 7

Heaths

Believe it or not, heathlands are a rarity. Not in Wales, perhaps, where most of us will remember picking bilberries, or whimberries as they are often called here in the west, on sunny July days, surrounded by the buzzing of insects and with the sward studded with the yellow flowers of tormentil and the white of heath bedstraw. Unfortunately, the insects usually include that pest of summer afternoons, the midge. Heaths are found in areas that could easily become woodland, but have been grazed just enough to prevent trees from covering the ground. They are defined as areas dominated by dwarf shrubs – mainly heather and gorse, but including a number of other species like bilberry, bell-heather and cross-leaved heath. They are valued for nature conservation because they are restricted to western Europe, and are rather rare on a world scale. They have a very limited number of plant species, but some interesting ones, including insectivorous plants like the butterwort and the sundew, which supplement their nutrition by catching and absorbing insects – a readily available source of nitrogen and phosphorous compounds. There are some unusual insects too, some of which, no doubt, end up as supper for a hungry sundew or butterwort! Knowledge of soil-plant relationships on this type of land is particularly useful, and of more than academic interest, as it enables you to plot routes across the land that avoid getting your feet wet.

Here I will discuss the more lowland heaths, generally below about 400 m altitude. Areas higher than this will be covered in the section on Moorlands. Lowland heaths in Wales are found in areas where grazing has been maintained at a low level, and where soils are acidic and generally shallow or peaty. Coastal areas on hard rocks are something of a special case: the conditions for heath vegetation are met because wind exposure severely curtails the growth of trees, whilst grazing is limited mainly to rabbits because stock is fenced off from the cliffs. Inland, heath develops on some common lands where, again, stocking rates are kept low. Brown earths are usually too fertile for heathland, but it can occur under exceptional circumstances, where naturally acidic land is very lightly grazed and fertilizers are excluded. More commonly, the lowland heaths are on rankers like the **Powys** and **Skiddaw** series (see p.117), or on **Manod** soils (p.88). The word heath can also apply to some wetlands, and wet heaths grade into bogs. In soil

Colourful heath vegetation near Barmouth

terms, it is probably best to think of peaty gley soils as supporting wet heaths in our westerly climate, whilst peat soils proper (with peaty horizons deeper than 40 cm) favour bog vegetation. Hence we would expect to find heather and cross-leaved heath on heaths, but not the bog species like bog asphodel and cotton-grass.

Oliver Rackham in *The History of the Countryside* gives evidence that heaths are mainly a man-made construct: " ...before 6000 BC the traditional heathland areas were covered by trees; heath vegetation existed, if at all, only in very small patches on cliffs and other special habitats." Heath is thought to have spread extensively in the Neolithic and Bronze Ages as grazing and burning became more common from about 1500 BC. Heath areas have often remained uncultivated, and as a result they are peppered with archaeological remains.

Thomas Hardy's famous novel, *The Return of the Native*, describes life on a Dorset heathland in the nineteenth century. Cutting bracken and gorse (furze) provided a useful addition to the food and bedding of cattle, whilst peat was cut for fuel, and briars (blackberry runners) were used as binding twine. He even records an early enclosure of the heath to create a copse of *firs and beech*. As techniques became available to increase the net productivity, mainly by the addition of synthetic fertilisers, so the traditional uses have declined. The main uses of heathland today are in nature conservation, sport and recreation, maintenance of grouse moors and the provision of open areas with high landscape value.

As an example, let me transport you to the Mynydd Bach, a few miles south of Aberystwyth. If you stand on one of its eminences, above Llyn Eiddwen, next to the beatifully carved memorial slate to four renowned local poets, you can look out westwards towards the sea, and across the fertile lowland of Ceredigion. It's a good place to look at the types of land that I have described in chapter 5. You are in the middle of a block of semi-upland covering some 15 square km, and rising to 361 m (1,190 ft) at Hafod Ithel. It is a kind of island in the middle of the fertile crescent of the Ceredigion plain. To the north lies the Ystwyth valley, cutting through the landscape; to the east is the Teifi, coursing lazily through Tregaron bog, with the higher uplands of the Cambrian Mountains beyond. To the south is yet another river, the Aeron, which rises on the south-eastern flanks of the Mynydd Bach itself and curls around its southern edge towards Aberaeron. You may be unfortunate enough to be in the middle of one of the Atlantic depressions which frequently sweep in from the west. The rising moist air from the sea soon drops a healthy proportion of its moisture on these hills: the rainfall here amounts to about 1,400 mm compared with 1,000 mm at Aberystwyth. Let's suppose, however, that it's a pleasant day (despite what I've said, they aren't rare), and you have a wander around the area. Between the rocky hillocks you will find several hollows and minor valleys, where the land looks significantly different. These basins are lined with till (*boulder clay*), which is also plastered around the flanks of the hills, and here the soils include large areas of peats and wet peaty gleys, (*stagnohumic gley soils* in the British classification). The intervening ridges are much drier, and have a few rock outcrops, and a scattering of Bronze Age tumuli which look like circles of stone overgrown with the heathy vegetation. The soils here are brown podzolic (Manod) and shallow soils (Powys) with some Stagnopodzols (Hafren) on the highest parts. Sorry – I know we haven't met the Stagnopodzols yet – they are described in chapter 11.

There are two attractive natural lakes near the

summit, Llyn Eiddwen and Llyn Fanod, both of which are nature reserves, surrounded by boggy wetlands. Most of the Mynydd Bach is used for extensive sheep farming on permanent grasslands and heathlands. This area accommodates one of the first wind farms in the Wales, a use of the land which, although controversial from the aesthetic point of view, is not incompatible, on the whole, with either farming or nature conservation. There is a mosaic of heathy and other types of vegetation, as well as the generally wet pastures: **heath vegetation** dominated by heather, gorse and bilberry; **acidic grasslands** responding to various combinations of factors, including drier soils, increased grazing pressure and small fertility differences, and **mires** – boggy areas where soils are slowly permeable, or where water tables come close to the surface. Patches of mire are usually determined by basin sites, or by the presence of springs. They occur close to the two lakes on Mynydd Bach, and in some hollows where ancient lakes have silted up. There are also a couple of conifer plantations.

Soils under heath vegetation vary from very dry, to wetter types. The dry ones are most common, usually sandy podzols or shallow soils over rock. Where profiles are wetter, for example over till, a proportion of mire species is found – like cross-leaved heath, rushes and *Sphagnum* moss. An intermediate vegetation, sometimes called bog heather moor occurs, and may include substantial amounts of cross-leaved heath with typical acidic mire species like cotton grasses and bog mosses. These mosaics of vegetation support a range of plants and animals that may not be found elsewhere, including many insects.

Nowadays we recognise that heath communities need to be protected from numerous threats, and we are trying to recover some of this type of land which has been used for more intensive types of agriculture or forestry. In recent years bracken has extended its cover over both acidic grasslands and heath. Control can be difficult, and the rhizomes are very difficult to eradicate by cutting. Effective specific herbicides have been available for several years. Another of the problems facing conservationists is the fragmentation of heath lands by drainage, farming and forestry. Drainage, surface cultivation and fertilisation are all agricultural improvement techniques which may need to be reversed in some way if semi-natural communities are to be restored. It is not enough just to leave the land to its own devices: this will usually result in invasion by scrub and trees. A certain amount of grazing is necessary to prevent succession to woodland, unless an area is above the climatic tree line, or subject to severe exposure from cold, or salt-laden winds.

For thousands of years, heaths have played an important role in human history: they have furnished resources for recreation, agriculture and homemaking, peat extraction and landscape. The Mynydd Bach, like many other open heaths, is criss-crossed with old rights of way, perhaps reflecting past use of common grazings and access

Heath and ffridd near Ponterwyd

for cottagers. Abandoned farmsteads and *hafodau* – shielings or upland shelters and dairies – are dotted about the plateau. Traces of prehistoric occupation are abundant too, in the form of the Bronze Age cairns on the drier parts of the hills with, no doubt, more archaeological remains yet to be discovered.

Excursion – Panorama Walk, Barmouth

OS map Outdoor Leisure 18 – Harlech, Porthmadog & Bala (SH625166)
Total distance – about a mile there and back; reasonably flat and easy, with a few rocky steps.

The car park for this walk is up the steep minor road that goes NNE from the eastern end of the town, passing the Bae Abermaw hotel. After about 1 km (0.6 miles) there is a free car park (SH625166), much used by the minibuses that transport climbers to the nearby crags. The route to Panorama Walk is well signposted on the left, after you go back down the road for less than 100 m. It is a short stroll of less than a mile to the superb viewpoint on the end of a steep spur overlooking the Mawddach estuary. The rocky land hereabouts is mapped as Revidge series – a very shallow, stony soil over hard sandstone.

This is ideal land for heath and, in the absence of heavy grazing, colourful mixtures of heather, bell heather, western gorse and bracken furnish an amazing foreground to the superb seascapes and mountain views from this spot especially from late August to September when the heathers are flowering (see picture on p.82).

If you want a longer walk, there are a number of possibilities. One of my favourites is to go up the lane that heads westwards from the car park and passes the climbing crags that are so popular with school and youth parties. Beyond these spectacular cliffs the little road arrives at the old farm of Gell-fawr where, behind the house, there is a horse-whim – a circular platform with geared machinery which was driven by a small horse pushing a bar round and round. These were quite common in rural Wales, and used for grinding grain and animal fodder. This one was used more recently to power a saw, and the mechanism can still be seen in the adjacent building.

From Gell-fawr a track leads south-south-west to Cell-fechan, and the old coach road that crossed the hills above Barmouth. Following this green road southwards brings you to a junction, where you turn right, and zig-zag past the big Catholic church into Barmouth. The old route from the town to Panorama Walk is marked by very old signposts here and there. It can be followed by climbing the steep steps past a couple of houses, at the eastern end of town, and skirting the edge of a large stone quarry. The path is rough and steep for a bit, but soon becomes a lane which passes some quaint cottages, and returns you to the car park.

In other parts of Wales there are some great walks visiting heathlands, for example:

- On the **Great Orme**, Llandudno, where there are rare limestone heaths – see chapter 10. You can take a tram ride from the town, or drive up to the car park on top of the limestone massif. If you keep a sharp look-out in August – September you may see the unusual autumn ladies' tresses with small white orchid flowers spiralling around the stem. It grows in the more grassy places.

- In the Rhinogydd, especially around the fabulous **Roman Steps,** which are accessed via the car park at Cwm Bychan (SH646315), the rocky terrain is ideal for heathland.
- Around **Llyn Barfog, Meirionnydd.** If you walk up to the lake from the car park on the Happy Valley (Cwm Maethlon) road, between Pennal (Cwrt) and Tywyn, at SN640986, you will see good examples of heathland, especially to the east and south of the lake. There is a good walk back to the car park along a track south of the lake, where there are excellent views over the Dyfi estuary. The round trip is about 3 miles, and the vicinity of Llyn Barfog (Bearded Lake) is ideal for a picnic.
- Along the **Pembrokeshire Coast Path** – especially around St David's Head, a walk of about 2 miles there and back from Whitesands Bay, which has a large car park and a pleasant café. You will also see the impressive Neolithic dolmen, Coetan Arthur, and, on the headland you may spot the remains of Iron Age hut circles.
- From Penmaen on the **Gower Peninsula** (SS527885) footpaths head north, or northwest onto the summit ridge of Cefn Bryn. The ridge is composed of hard conglomerate (pebbly) rocks of the Devonian system. Here there are rather unusual soils for Wales – humo-ferric podzols – which occur just on these hard, porous rocks. There is a mixture of dry and wet heaths, which you can observe by walking west-north-westwards along the ridge to the viewpoint beside the road at SS494899. If you feel like it, you can visit the burial chamber, Arthur's stone, about 500 m northwest of the road, and return the way you came to Penmaen, in which case you will have walked about 11 km (7 miles).

CHAPTER 8

Ffridd Lands – The Lower Hills

Above the spring lines that have supported settlements and farms for millennia, there is usually a steep hill that is useful for extensive grazing of cattle or sheep. Where this land has been enclosed with large fields it has been called ffridd (pronounced *freethe*). The word must be an ancient one, as it has found its way into so many place names. The ffridd is the enclosed land above the smaller fields and pastures around the farm, but below the mountain wall which separates the farmland from the open mountain or moorland. This type of land occupies large parts of Wales between about 200 and 450 m above sea level. Over extensive areas it is dominated by one soil series, known as **Manod**, a Brown Podzolic soil which shows up as broad areas of pink on the National Soil Map. Interestingly, *manod* means fine snow, a not altogether inappropriate name for the soil, as this land can be powdered with snow when the lower lands are still green. The name, incidentally, comes from Manod Mawr, a large rounded hill near Blaenau Ffestiniog (see photo, p.24), in whose slate tunnels some of our national treasures were hidden during the Second World War. The single most striking feature of Manod profiles is the rich orange-brown colour of the subsoil, which is developed from the weathered mudstones and siltstones stained with rusty-coloured iron compounds.

The consistently sloping nature of this land has had a profound effect on the soil development. Millimetre by millimetre over the millennia, the upper layer of topsoil is continually creeping down the slope, and, often, being replenished by weathering on the upper slopes and plateaux. As iron-rich particles of soil move across the land, the rain washes iron (and other minerals) out of them and deposits it lower down in the soil profile. The iron accumulates at depths of between 20 and 50 cm, and, because the soils are permeable, the iron is oxidised (i.e. rusted) in the presence of air in soil pores to give the characteristic reddish brown colours. The iron has other effects too, both chemical and physical: the soil aggregates or peds tend to be coated with iron-rich material which acts as a loose cement and stabilises them, giving a rather fluffy, pellety structure to the subsoil. This is so pronounced that people often think the subsoil is sandy, as the tiny peds can look like coarse sand particles. In fact the peds will rub down to silty clay loam – quite a fine texture – as the original

Ffridd lands near Kington, Radnorshire

soil particles are rich in clay and silt derived from the dominant parent mudstones. Chemically these soils are acidic, and lacking in many plant nutrients, especially phosphorus and nitrogen. The addition of these elements as fertilisers will 'paint' the grasslands a bright green, an effect that is often very striking, especially on bright spring days after rain. (The effect has been used occasionally to literally paint the pastures with a message, usually of goodwill to newlyweds.)

Unfertilised, the soils can support acidic grasslands rich in moderately nutritious grasses like bents and fescues. Unfortunately for the farmers, these types of land are particularly susceptible to invasion by bracken, which shades out many other plants and covers large areas where the soil is deep enough for the rhizomes to be well protected from frost. It is a common sight to see the lower, deeper soils of the ffridd covered with bracken, giving way to gorse on the upper, convex slopes where soils are shallower over rock. Bracken is so common in Wales that the Welsh word for it, *rhedyn*, is also the general word for fern.

It is the Manod soil which inspired the title of this book. It is pleasingly appropriate (and, perhaps, not an accident) that the deep bracken soils which were valuable enough in the context of Welsh hill farming to inspire the proverb, do, in fact, have golden subsoils!

In a few places the perniciously invasive rhododendron, mentioned in the context of woodlands above, has spread from Victorian plantings, and covers hillsides in a spectacular show of colour for a few weeks in summer. A blowsy thicket of this type can be seen at Dinas Mawddwy on the west-facing slopes of the Dyfi, opposite Meirion Mills Woollen Centre.

In the absence of competitive weeds like bracken and rhododendron, the pastures can develop species-rich grasslands, especially if the acidity is not severe. The more interesting and varied grasses and herbs are not tolerant of strong competition from agricultural grasses and clovers, but will thrive in pastures of low fertility where the grazing is persistent but not intensive. Traditional hay meadows fit this description, and some of the higher altitude farms, and those close to the coast where oceanic climates give cool, rainy summers have enclosed hay fields that have been unploughed for long periods. They may have been limed in the past, which has reduced acidity somewhat, but have had little artificial fertiliser. The combination of these factors can result in flower-rich swards which are a delight to the eye, and of great interest to conservationists and ecologists. They can often be spotted from afar by the presence of colourful flowers like the yellow cat's ear (a dandelion-like plant), ox-eye daisies or yellow rattle. Some may have orchids growing amongst a variety of colourful plants like red clover, betony, bird's foot trefoil, vetches and knapweed. Such meadows have been rare since the 1950s, but are making a comeback with the encouragement of government agencies like DEFRA and the Countryside Council.

The location of the mountain wall can give clues about the nature of past settlement, and its position is of particular interest to archaeologists. For example, the Cistercian brothers at Strata Florida Abbey, Pontrhydfendigaid, Ceredigion, encouraged sheep farming to extend further into the hills, and

Species-rich meadow on ffridd land in Powys

Ancient ffridd wall, Strata Florida

the mountain wall winds its way along the contours between 320 and 360 m. The variations relate to the nature of the land, with patches of less rocky and stony land being included in the ffridd, whilst the wall may descend if there are difficult patches at lower altitudes. If there is a kink in the wall, it may indicate an ancient structure around which the wall deviated.

The ffridd lands are by no means homogeneous, and although Manod soils are prevalent on them, many other soils are found. Shallow ranker soils occur on convex hummocks and towards the summits of the ridges, and there are occasional rock outcrops. Wetter soils frequently break up the pattern, especially on lower slopes, where till has been deposited (see next chapter). At the somewhat higher altitudes of the ffridd, the wetter soils tend to have humose or peaty topsoils, and are often affected by springs or seepage water coming from the bedrock, or flowing at the junction of bedrock and till. Some of these have been mapped as peaty Surface-water Gley soils, **Wilcocks** series, especially on the National Soil map, as the groundwater affected areas (**Freni** series) are usually too small to map at the 1:250,000 scale. It is also rather difficult to differentiate in the field between surface water, that is held up by very slowly permeable subsoils, and groundwater that is seeping into the soil from the sides or from below. The wetter areas can be readily seen, as they are often rushy, and irregular in tone and fine scale surface topography compared with the even shades of the Manod soils. They tend to form elongated isosceles triangular shapes with the broadest part towards the top of the slope, narrowing down into a stream towards the bottom of the hill.

Not all of the ffridd zones are on the acidic mudstones and siltstones that give rise to the Manod series. On the Old Red Sandstones which covers large parts of south Wales, for example, there are reddish well-drained soils known as **Milford** series (see pp.59–60). Similar climatic zonation occurs, expressed by leaching of iron at higher altitudes, but the naturally red colouration (also caused by oxides of iron in these Devonian rocks which were formed under desert conditions) tends to mask the accumulation of oxides in the subsoils, whilst the higher natural fertility of these soils reduces acidity and the tendency for podzolic horizons to form. Hence the Milford brown earths of the lowlands tend to extend into the ffridd zone, and the mountain walls tend to be higher in the landscape. In Snowdonia the mountain wall is at lower altitudes in the west because the oceanic effect (see p.12) is stronger there. It is also lower where hard rocky outcrops made ploughing difficult or impossible in the days when every farm had some arable fields. Similarly on the scattered limestone areas, the location of the ffridd is controlled by geological factors, usually the presence of extensive limestone crags and outcrops.

Excursion – Strata Florida – Ystrad-fflur

OS map Explorer 187 – Llandovery / Llanymddyfri (SH745657)
Total distance – 7 miles; hill paths with some boggy parts.

Ystrad-fflur is one of the iconic sites of Wales, and well worth a visit to see the Cistercian Abbey and its picturesque surroundings. Archaeologists from Lampeter University are doing a lot of work, and beginning to amass evidence of the life and economy of the middle ages in this region. The ruins of ancient farmsteads are scattered over the hills, especially in the vicinity of an antique ffridd wall.

There is a waymarked trail from the abbey – the *Teifi Pools walk,* and this walk will provide access to much of the landscapes and soils described in chapter 11. However, a shorter excursion, of about 7 miles enables you to appreciate the differences between the enclosed farmlands around the abbey, and the ffridd lands, where there are romantic and eloquent traces of a forgotten way of life. Follow the waymarked Teifi Pools walk (it is shown on the Explorer map No. 187) eastwards up the lane to Tyncwm (SN771655), where you turn left up the farm lane, and follow signs which take you onto the hillside above Nant Egnant. The path goes ENE, and crosses a shallow ford (which can be difficult after heavy rain). The way continues in the same line, climbing, via a rather boggy patch, where you can practise the Laws of Bogology (see p.54) towards Llyn Egnant, which is worth a visit. A little before the lake, however, there is a large tumulus – a rounded pile of stones with an earthen rampart – this is a Bronze Age burial site called *Banc Blaenegnant round cairn.* Over the years, passing wayfarers have added stones to the top, to improve its function as a waymarker, so the cairn can truly be said to have a continuing use and significance. Here you can turn sharply back on a fainter footpath that heads due west, past a sheepfold and then across a stream. Although this path is faint, it is quite evident, and, where it crosses streams and wet flushes, there are strategically placed stones that enable you to pass dry-shod.

You have been in ffridd lands since crossing the ford, and the path, after turning slightly to the north, descends through the ancient ffridd wall to the farm of Frongoch (SN766667), from where an access track goes SSW to Troed-y-rhiw, and on to the metalled road, where you turn right, to pass the water works and return to the abbey.

Any walk that involves climbing onto the hills and moorlands will lead you through these ffridd pastures, so keep an eye out for them, and for the archaeological remains, some of which are not shown on the map, that are often to be seen in these little cultivated regions. They are often haunted by skylarks, whose trilling song is a welcome accompaniment in areas where other people may be very thin on the ground.

CHAPTER 9

Lowland bogs and rhosydd – seepage bogs

Peat occurs sporadically across our area, but always leads to interesting habitats. I will deal with the large areas of upland and blanket peat in a subsequent chapter, but here we will look at two types of peat found in lowland situations: raised bogs and valley bogs.

Peat is a particular type of soil that most people recognise, although in different guises. Perhaps the most frequent way in which it used to be encountered was in bags at the garden centre, where it was sold as a ready source of organic matter to improve the water holding capacity of garden soils, and for use as a mulch and fertiliser or growth medium, the latter being in special composts, mixed for seedbeds or for mature plants. Generations of gardeners were familiar with the John Innes range of peat-based composts. They were dark-brown, dry, crumbly preparations made using peat sourced in raised bogs, mainly in Britain and Ireland (and some from Russia, I believe). This use of raised-bog peat is frowned upon nowadays as the stripping of bogs is non-renewable, and damages valuable wildlife habitats. This is in real contrast to domestic-scale peat-cutting for fuel, which was an important element of household economy in many parishes, and went on well into the last century on most of our bogs. It was an activity that actually *increased* biodiversity, as it provided a range of habitats in the bogs, including shallow pools and drier banks. The rate of extraction usually did not exceed the powers of the bog ecosystem to regrow the peat. The shapes of several parishes may well have been influenced by the need to include peaty land – a turbary – within the parish boundaries. Upland bogs were used by many parishes, but it was a bonus for the people if there was a more lowland source of peat. The climate at lower altitudes is much more favourable for the important business of drying the peat. Special tools were used to cut long bricks of peat, which were stacked in an exposed spot, near to the source, in a lattice pattern that maximised the circulation of air around each brick. As you will appreciate, it is much easier to transport light, dried bricks of relatively clean peat than to try to carry it home in its raw state, heavy with filthy water that oozes over everything in the vicinity.

The other guise in which peat, or turf, as it is sometimes called, is encountered, is in its natural state in boggy land, where waterlogging slows the rate of decomposition of wetland plants, causing an organic-rich soil to build up over hundreds, and often

Tregaron Bog

thousands of years. Peats like this are seen mainly in upland areas, where the cool summer climates and high rainfall promote surface accumulation of acidic, rotted organic matter. However, there are several places in the west of Wales, and a few further east, where waterlogged places still carry large quantities of peat. One of the main causes of waterlogging is the presence of impermeable subsoils which prevent the penetration of surface water, mainly rainwater, through the soil. In most parts of Wales there are large areas of thick boulder clay, more properly known as till. The material is composed of mixed deposits laid down by the glaciers as they rode across the country in a number of separate glaciations during the last 2.4 million years. The powerful sole of the glacier ice scraped and ground the local rocks into a fine paste forming a heavy clay loam. At the same time, the ice sheets had carried 'foreign' rocks, sometimes from far afield. This material was plastered over the valley bottoms and sides, and sometimes over more level plateaux as the glaciers made their inexorable progress over the land. Geological maps, known as *drift maps* show the distribution of this material, and these maps are in good accordance with the soil maps, which, in effect, show the same thing by mapping soils with till substrates. These are mainly the Surface-water Gley soils incorporating Cegin and Brickfield series (Stagnogley soils), and Wilcocks series (a stagnohumic gley soil).

Where the boulder clay is easily drained by

artificial means it provides adequate pasture soils, but it is usually too wet, in our rainy climate, for intensive arable use. Some areas are in low-lying positions, affected by springs, or on hummocky ground that it is not economic to drain. These have remained as rough grasslands, and often provide useful areas for wildlife. Hollows in the till plains, and areas where valleys have been blocked by accumulations of glacial debris, or by landslips, have peaty soils of various kinds. We have discussed some of these, called peaty gley soils (more properly stagnohumic gley soils in the current soil classification), in Chapter 5. Where the surface peat layer is deeper than 40 cm, the soil is classified as a ***peat soil***.

If you study the 1:250,000 soil map of Wales you will see extensive peat soils, coloured purple, covering some of the highest land, with only a few splodges in the lowlands. The most notable of these latter are the raised bogs in Ceredigion: Borth bog in the north of the county, and Tregaron bog in the centre. There are a few significant areas of valley bog in Anglesey, and a large area, Fenn's Moss, at the eastern extremity of Wales, straddling the border into Cheshire. In south Wales, there are mappable (at the national map scale) areas of peat at Crymlyn bog between Swansea and Neath; behind Oxwich on the Gower peninsula, and at Gors Goch, Llanllwch, west of Carmarthen. This is only half of the story, however, since many bogs are too small to appear on the national map, which cannot show areas of less than 50 hectares, i.e. half a square kilometre. A look at some of the more detailed soil maps that cover about 20 per cent of Wales will reveal many more small bogs and valley mires. Ecologists have made an art form out of classifying different types of peat lands according to the chemical and physical nature of the peat, and its hydrology, and the types of vegetation that grow on it. Here is not the place to go into detail. The most important aspect of the soil classification is the separation of raw peats from earthy peats. The distinction recognises that, although all peats have formed from the slow decomposition of plant material under waterlogged conditions, some have subsequently dried out, usually with the help of artificial drainage, and have been cultivated. This process, over the years, produces a topsoil with some of the characteristics of mineral soils: the development of an earthy structure. Other areas are a mixture of organic peat with mineral components that have been mixed in by various processes, mainly by washing in with spring or river water from adjacent hillsides.

Sedge-rich pastures or *rhosydd* are a characteristic feature of west Wales, on particular types of wetter land. The Welsh name *rhos* has come to be used nationally as a name for this type of vegetation. Eighteenth-century surveyors working in Wales used "rossy" as an adjective for this kind of terrain. It is a type of poor fen of the Atlantic regions – the climatic area stretching from south-west Norway, via Denmark, the Netherlands and western France to north-west Spain and Portugal.

Pingo at Rhos Glynyrhelig, Gorsgoch

The Atlantic climate is distinctive in its cool, moist, cloudy nature. Temperature fluctuations are relatively small from the seasonal to the daily scales and the mean yearly temperature range is less than in any other part of Europe. Rainfall is moderately high throughout the year, and evaporation generally restricted by the moist atmosphere, so that there are relatively good soil moisture supplies well into the summer. In the rhos pastures, the supply of soil water is augmented by seepage water from springs, or high groundwater, so that they are usually wet for most of the year. The seepage waters carry nutrients from their catchment areas, but acidic peat forms a surface layer to the soils, so they have acidic topsoils, but some nutrient enrichment in the subsoil. Sedges are particularly well adapted to cope with waterlogging, but the grasses, in general, are not (with the notable exceptions of reed (*Phragmites australis*) and purple moor-grass (*Molinia caerulea*).

Raised bogs, also called raised mires, are decidedly different. For a variety of reasons the growth of the peat has raised the soil surface above the water table, and the upper parts are therefore watered only by rain and capillary action, leading to very low nutrient levels. The peat is rather different in nature from other peats, being derived largely from species of the moss *Sphagnum*. It is almost permanently waterlogged and has a rather fibrous nature, so that when you squeeze a mass of it in your hand, the water that comes out is almost clear. The vegetation on these bogs is low and heath-like, dominated by mosses and low-growing herbs and shrubs; although recently willow scrub has been invading the fringes of most of our bogs. They are notoriously wet and soft, so that if your feet break through the surface vegetation you are in danger of sinking right in. (Naturally, I have personal experience of this – it's no joke if you are on your own!) The larger raised bogs have an appreciable domed look to them, as the middle of the bog can be some 5 metres above the surrounds.

These peat lands are of little use to agriculture unless they are drained and the peat is allowed to dry out. Deep fen peats in the east of England have been treated in this way, and produce excellent soils when fertilized and carefully managed. In Wales, however, we have been spared the ennui of such agricultural plains because the wet climate and the topography of our bogs makes reclaiming them for agriculture just too difficult and expensive. We have two particularly good raised bogs in Wales, both in Ceredigion. Cors Caron, or Tregaron Bog, covers some 768 ha (1,900 acres). It is an internationally listed wetland of prime ecological and geomorphological importance. It has formed on the moraine-dammed flood-plain of the river Teifi (a *moraine* is a pile of stony debris deposited by a melting glacier). Borth Bog, or Cors Fochno is a somewhat similar bog, also of international importance, but influenced by its proximity to the sea. These are unique habitats with several rarities in both the plant and animal kingdoms, but there are also plenty of typical bog plants like the bog asphodel, bog bean and cotton grasses as well as the rarer ones like the long-leaved sundew.

Attractive hide on Tregaron Bog

A typical rhos area in Ceredigion

Rhos Glynyrhelyg (*Glyn* = valley, *Helyg* = willow), is a good example of unimproved sedge-rich pasture on wet boulder clay soils. It lies along the river Grannell, a tributary of the Teifi in Ceredigion. The rhos occupies about 16 hectares about a mile north of the small village of Gors-goch, which, incidentally, boasts another nature reserve on the valley bog that gives it its name (*Gors* = bog, *goch* = red). The bog covers about 40 hectares south of the village, and was an important source of fuel for the villagers in the past. Many Welsh bogs contain the element *coch* or *goch* because the dominant sedges of these deep peats have reddish stems and leaves – especially in autumn and winter.

In Britain some of the characteristic plants of the rhos, or poor fen community are:

sharp-flowered rush (*Juncus acutiflorus*)
jointed rush (*Juncus articulatus*)
bulbous rush (*Juncus bulbosus*)
meadow buttercup (*Ranunculus acris*)
lesser spearwort (*Ranunculus flammula*)
marsh willow herb (*Epilobium palustre*)
sneezewort (*Achillea ptarmica*)
marsh hawk's beard (*Crepis paludosa*)
carnation sedge (*Carex panicea*)
common sedge (*Carex nigra*)
star sedge (*Carex echinata*)
marsh pennywort (*Hydrocotyle vulgaris*)
marsh lousewort (*Pedicularis palustris*)
water mint (*Mentha aquatica*)

In addition, the Welsh rhosydd, of which Glynyrhelyg is typical, have much purple moor-grass (*Molinia caerulea*) and soft rush (*Juncus effusus*), with the local and unusual whorled caraway (*Carum verticillatum*) – a member of the carrot family and the county plant of Carmarthenshire – and Devil's bit scabious (*Succisa pratensis*), the important food plant of the declining marsh fritillary butterfly. Heath spotted (*Dactylorhiza maculata*) and lesser butterfly (*Plantathera bifolia*) orchids are sometimes found and can be impressively abundant, as they are in parts of Rhos Glynyrhelyg.

This site is important as a representative of the rhos flora, but it also has some special features that enhance its conservation value. One of these is the presence here of post-glacial features called **pingos** developed in the boulder clay of the area. Pingos were formed during the period immediately after the ice from the last glaciation began to melt about 12,000 years ago. Lumps of ice the size of a bus, or even a large house, breaking off glaciers tended to become buried in the boulder clay forming below the ice. When the glaciers eventually melted, the buried ice remained for some time before melting to form depressions in the boulder clay known as kettle holes. Sometimes, however, the ice remained insulated below the surface, where it retained its very low temperature. Seasonal melting, and other more complex phenomena associated with the presence of deep permafrost, allowed surface water to seep down cracks, or to be squeezed out of soil pores and freeze onto the buried ice, causing it to grow. As successive seasons provided additional seasonal growth, thus expanding the ice core, it heaved and compacted the surrounding soil, and eventually burst out onto the surface – these features can currently be seen in Arctic/Antarctic tundra regions. Eventually, as the climate became warmer, the ice melted completely, leaving a large water-filled hole, surrounded by a ridge, or rampart of compacted soil, so that the feature looked like a crater. They vary in size, but our Welsh examples tend to be 20 to 40 m across. Some pingos still contain pools, whilst others filled with peat as the pool was colonised by plants – a process familiar to ecologists, called succession. Kettle hole and pingo topography provides a range of niches for a variety of plants. For economic farming or forestry, such features are a nuisance, but they have served to preserve some wild areas for us.

Drier areas at Glynyrhelyg have herb-rich bent-fescue pasture with species like cat's ear (*Hypochaeris radicata*), knapweed (*Centauria nigra*), eyebright (*Euphrasia spp.*) and the unusual dyer's greenweed (*Genista tinctoria*). Other rarities include the small fern, moonwort (*Botrychium lunaria*) and the lichen *Usnia articulata*, which grows on stunted hawthorns. The ridges around and between the pingos have dense subsoils, and tend to be droughty in summer: they are being invaded by gorse, and heather (*Calluna vulgaris*) occurs in a few places.

The pingo hollows contain waterlogged peat soils bearing ***rich mire*** communities. Mires are plant communities on more or less permanently waterlogged soils. Here they are dominated by the Sphagnum mosses, and with typical bog species like cross-leaved heath (*Erica tetralix*), bog bean (*Menyanthes trifoliata*) and bog asphodel (*Narthecium ossifragum*), but also some more unusual species indicative of a higher nutrient status, like marsh cinquefoil (*Potentilla palustris*) and bottle sedge (*Carex rostrata*). These basin mires are unlike the raised bogs (e.g. at Borth and Tregaron) in that they receive water from the surrounding slopes and the groundwater table, which tends to be flowing towards the river. The minerals in the basins are thus being constantly renewed at a slow, but steady rate.

Some rare insects are associated with the mires, and include the small red damsel fly and the keeled skimmer. The more common, but attractive banded demoiselle can be seen flying over the river.

The reserve is estimated to have some 120 species of flowering plants, 18 species of butterfly, and numerous other insects, including the rare forester moth.

Management of these rhos areas depends on the maintenance of the high water tables and the continuation of light grazing to prevent succession to carr and woodland. Their presence in our area owes as much to the common grazing practices of former years as it does to the soil, climate and glacial history of the region.

Excursion – Tregaron Bog – Cors Caron

OS Explorer map 199 – Lampeter / Llanbedr Pont Steffan (SN692624)
Total distance (to the hide and back) – less than 1.5 miles; flat and easy.

If you drive up the B4343 from Tregaron northwards towards Pontrhydfendigaid, within a couple of miles you will arrive at a car park (on the left), provided by the Countryside Council for Wales. Here there are information panels, and you can visit the level, wide boardwalks which give access to the famous raised bog. There is an excellent (and beautifully designed) hide overlooking the Teifi, less than a mile from the car park. On the way you will see the typical raised bog vegetation and soils that I have described, as well as some very pleasant stretches of open water where swans, ducks and geese can be seen feeding and nesting. Or you can take a walk north-eastwards along the long abandoned Aberystwyth to Carmarthen railway line. Some short boardwalks extend into the bog from this route, and there are ample opportunities to appreciate the distinctive flora and fauna. The boardwalks have been designed with wheelchair users in mind, and this is also an excellent place for families with young children to visit (as long as they stay on the boardwalks). Tregaron bog, with its delightful access arrangements, provides an almost unique opportunity for all of us to experience a very special type of habitat in comfort and safety.

Many lowland bogs, rhosydd and fens throughout Wales are also nature reserves where visiting is encouraged, and most have facilities to enhance your understanding and enjoyment of the sites. The raised bog at Borth is used more for research purposes, and can be visited by permission from the Countryside Council for Wales. You can visit the fringes, however, by using the new permissive path, part of the Ceredigion coast path, from Borth to Taliesin.

CHAPTER 10

Limestone landscapes

Limestone landscapes are always distinctive, indeed, some of Europe's most dramatic scenery is formed by these hard, pale rocks, including the magnificent jagged peaks of the Dolomites in Italy and the Picos de Europa in northern Spain. It is a matter for pondering that large areas of the world are on limestone rocks, some of them hundreds of metres thick – and much of this rock is the result of living things that scavenged dispersed calcium ions from the seas and concentrated them into their shells. So the very fabric of the earth's surface has been engineered by the processes of life, long before we humans appeared on the scene.

Our home-grown landscapes are gentler than many in the world, but certainly have their own grandeur, as the impressive scarp of the Eglwyseg in Powys attests: it is something of a test of nerve to walk along the foot of this extended cliff on the section of the Offa's Dyke path from Llangollen to the wonderfully named World's End, because, below the foot of the cliffs, the screes still fall away at a rather daunting angle.

The attraction of these landscapes is often in the wildness and the variety of small pastures, woodlands and scrub, and the diversity of the specialised flora, which has led to the creation of many nature reserves in limestone country. The topography is varied too, including rocky outcrops – like the one on which the evocative Carreg Cennen castle stands in Carmarthenshire – and the presence of old quarries, providing artificial cliffs and other niches for unusual plants. The unique nature of the drainage on these permeable and ultimately soluble rocks creates bizarre features, like the sinkholes in the wild country north of Abercraf in Powys. In fact many of our most unusual and exciting landscapes are on this type of rock, and some of the biggest cave systems in Europe are found below the moorlands of the south Wales limestones, where the steep scarp of the limestone is topped by a ridge of rolling hillocks with occasional sinkholes (shake holes or swallow holes formed naturally by dissolution of the limestone). It is dominated by old quarries and their spoil, and by frequent natural rock outcrops, some with scrub or copses on them. Most of the plants are stress tolerators, adapted to drought and grazing pressures, but ruderals (i.e. weed species) occur, as might be expected, in areas which are disturbed by sheep, rabbits and people, especially along tracks and by

The Eglwyseg scarp on Offa's Dyke Long Distance Path

gateways, where bare soil enables wind-blown seeds to gain a brief roothold.

Strictly, perhaps, this section should include most of the Vale of Glamorgan, where the soils are developed on carboniferous and lower lias limestones. However, the influence of intensive farming and the presence of large amounts of glacial deposits make this area unique in Wales, so I have

Carreg Cennen on its limestone crag

included it at the end of chapter 5 on lowland pastures and woods (p.68).

Wales lacks large areas of limestone such as those in Derbyshire and Yorkshire, and the extensive chalk lands of southern England, but this makes the areas where they are found all the more interesting. Our most northerly limestone is the Carboniferous on the east coast of Anglesey around Benllech. Much of this is masked by glacial deposits, although there are outcrops, and these become more obvious around Penmon and on Puffin Island. Another band of Carboniferous limestone (also known as the mountain limestone) outcrops between the millstone grit of Clwyd and Shropshire and the Silurian rocks of the Clwydian Hills and Powys. Dramatic hills at Llynclys and Llanymynech straddle the Powys-Shropshire border near Pant, representing its southern limit. You can usually pick out our limestone areas by the presence of quarries – old and new – for example from Dyserth and Tremeirchion in the north, via the eastwards displacement caused by the Bala fault, to Llangollen, Oswestry and Pant. Travelling south, there is then a large area without limestone until we reach the narrow strip which surrounds the south Wales coalfield. From Kidwelly, near the coast, it heads eastwards to the Llandeilo area (see picture of Carreg Cennen, left), and on, south of the Brecon Beacons to the area south of Abergavenny. Here it turns southwards following the west bank of the Usk via Abertillery and Caerffili into the Vale of Glamorgan. It can then be followed westwards along the southern flank of the coalfield, until it reaches the sea at Porthcawl. We find extensive limestone landscapes once again in the Gower peninsula, where spectacular cliffs and caves can be explored on the south coast, and around Burry Holms in the west. There are similar outcrops in south Pembrokeshire, especially from Stackpole around the coast to Freshwater West. In several areas the limestone landscape is moulded into the typical *karst* forms (named after an area in the Balkans). These landforms are the result of erosion of the limestones through dissolution. The naturally acidic rain, drip by drip, dissolves away the calcium carbonate, leaving characteristic shapes. The process can be simulated by dripping hot water onto ice. One characteristic of karst is a cracked surface consisting of deep linear gullies, called *grykes*, which separate knobbly blocks of weathered rock, the *clints*. In less enlightened times these clints were ripped up and used as ornamental features on walls and rockeries, and can still be seen aplenty in limestone areas. Another feature is conical hollows of various sizes, known as swallow holes, sinkholes or *dolines*. Water drains into these hollows and disappears underground, where it continues its work, reinforced by acids that leach from the soil, and creates cave systems and underground streams. In places, streams appear spectacularly from caves, as in the source of the Llwchwr south of Carreg Cennen.

The deep, moist, shady grykes of the limestone pavement are usually deeper than an arm's length, and half conceal interesting plants like bloody cranesbill, and the rare Hutchinsia (*Hornungia petraea*), as well as many unusual orchids including twayblade, bee, early-purple, fly and small-white orchid, and dark-red helleborine; whilst the true wild cotoneaster occurs in the UK only on the Great Orme. Other rarities found on the Great Orme limestones include Nottingham catchfly, goldilocks and horseshoe vetch. Areas left ungrazed at relatively low altitudes may turn into woodland, where plants like herb paris, nettle-leaved bellflower, and limestone woundwort grow beneath ash and hazel canopies, often with sycamore and elder. There are other rarities on the south Gower coast, described by Jonathan Mullard in the magazine *British Wildlife*, Vol. 20 (December 2008) as "one of the classic British botanical localities and a refuge for many species that are scarce or rare at a national level." Here you may find the hoary rock-rose and the goldilocks aster, as well as the yellow whitlow-grass, which occurs nowhere else in the UK.

The most notable soil on the limestones is called a **rendzina**. This is one of the most interesting of the soil names, as it is originally a Polish word, once, if not still, used in rural communities to signify the clinking sound made by the plough passing over stony ground. (Our phonetic pronunciation of the word, no doubt, does no justice to the Polish rendering!) This delightfully onomatopoeic word is used for a class of shallow soils over limestone, in which the topsoil is calcareous – i.e. it contains free calcium carbonate, which gives a fizz if dilute acid is dripped onto a soil sample. This makes for a very distinctive soil – see the section on pH on p.64. The rendzinas in Wales constitute a simple A–C profile, a dark topsoil immediately over shattered limestone, called **Marian** series.

Because the soils are shallow and stony, and the limestone is inherently infertile, there is much semi-natural vegetation to be found in these areas. The plant communities are distinctive, with much scrub and calcareous grassland, often with unusual, or even rare species. Plants relatively unusual in Wales, like bloody cranesbill, yellow wort, or viper's bugloss can be considered as indicators of calcareous soils in any part of Wales, whilst on the Gower grasslands and scrubs, for example, there are some more uncommon species like the hoary rock-rose, and spring cinquefoil, which also occur in Anglesey, on the Great Orme and a few other places in north

Wales. In a few places there are some national and international rarities, like a number of whitebeam (*Sorbus*) species which grow only on the limestone cliffs of Breconshire. Ley's whitebeam (*Sorbus leyana*) grows exclusively on the cliffs north of the coalfield, near Crickhowell.

The soil survey maps have given precedence to another soil called **Crwbin** series, in which the calcium carbonate has been washed out of the topsoil. Although there is still plenty of carbonate affecting plant growth, these soils are classed as rankers rather than rendzinas. Hence the map units with shallow soils over limestone are shown on the National Soil Map as "Crwbin Association".

As well as bare rock, limestone screes, rankers and rendzinas, there are some deeper soils over the limestone in Wales. Nearly all of these are leached, which is to say that the passage of rainfall through the permeable soil has dissolved the carbonates from the surface horizons, leaving a slightly acid soil similar to the Denbigh series – it is called **Malham** series, as it is over limestone rather than shales and sandstones – I have already discussed this soil in the section on the Vale of Glamorgan (p.68). The Malham soils have fine silty textures, and are usually associated with deeper soils in which the bedrock is below 80 cm – these are called **East Keswick** series. There is some controversy as to where the bulk of these deeper soils has come from. Limestone is a relatively homogeneous rock composed of calcium carbonate and little else (sometimes there is a component of magnesium carbonate). It weathers quite rapidly, as it is dissolved by slightly acid water, and rain is slightly acid, having picked up carbon dioxide on its passage through the air. The reaction between rainwater and the carbonates in limestone leaves very little solid residue, as most of the dissolved carbonate is carried away (sometimes to be deposited elsewhere as *tufa*). It is thought that a significant component of the soil material over limestone has been deposited by wind, which accounts for its high silt content. This, however, begs the question as to why wind-blown dust should land only on limestones! Perhaps there is a significant component of silty wind-blown particles in most soils, but where the soil is developed in non-limestone, the various mixed weathering products of the mudstones, siltstones or sandstones mask its presence.

Where Malham and East Keswick soil associations are mapped, it is usually fairly easy to predict soil depth from the topography, as the shallower soils are on convex sites like hill crests, and the deepest ones are on concave footslopes. These deeper soils on the limestone are almost exclusively leached, and their pH is around 6 to 7 – in marked contrast to the soils that we have seen on more acid rocks, whose pH is much lower. The rolling landscapes of the deeper limestone soils are notable for the lack of streams and rivers, as the rainfall in these areas tends to sink into the ground and then pass rapidly through the permeable limestone bedrock into underground streams.

CHAPTER 11

Moorlands and upland bogs

The very name Wales or Cymru conjures up an image of rolling moorlands, perhaps dissected by green valleys and backed by craggy mountains. It is difficult to define moorland, as it is an amalgam of land use, topography and elevation. In that fine book, *The Encyclopaedia of Ecology and Environmental Management,* edited by Peter Calow (1998), moorland is defined by P D Moore as "upland areas of heather-dominated vegetation in north-western Europe", and he goes on to discuss differences between heathland and moorland, and the fact that they are both man-made habitats, as they rely on burning and grazing to prevent tree growth, and an ecological succession to forest. Most of our upland areas have been heavily grazed by sheep for many years, or have been afforested with conifers, mainly Sitka spruce. I will consider the afforested areas later; for now let's deal with the traditional open moorlands where (nowadays) flocks of sheep graze on semi-natural vegetation. Several parts of Wales are famed for their moorlands, and some are given names like the Denbigh moors, east of Snowdonia, which incorporate Mynydd Hiraethog and Clocaenog Forest. More often the moorlands are called *mynydd* – like the several places called Mynydd Bach, or like Mynydd Eppynt, that large area of land between Builth and Llandovery that rises to 475 m above Llangammarch Wells. Several of the moorlands that separate the south Wales valleys are also called mynydd, like Mynydd Maesteg and Mynydd Merthyr. Mynydd Preseli, part of the Pembrokeshire Coast National Park, is a kind of cross between mountain (see next chapter) and moorland, having areas of rolling high country with several igneous crags sticking out of the tops.

Above the *ffridd* zone, which is discussed in chapter 8, the brown podzolic, mainly Manod soils begin to give way to much poorer, more peaty soils. Where the parent material is rock, usually on steep or moderate slopes, ***podzols*** are formed. You may have heard of this type of soil, perhaps from vaguely remembered geography lessons. The word, like so many soil terms is of Russian origin, meaning something like 'ashes underneath'. This gives us an entry into the nature of the beast. I think the name originated from people digging, or ploughing the land, and finding that a starkly contrasting white, or very pale earth appeared underneath the dark, humus-rich topsoil. This pale material has lost its colour as a result of **leaching**, a process whereby

Moorland on the Cambrian Mountains

the flow of water through the soil profile washes out the minerals that give it colour. We have already established that the main source of colour in subsoils is iron, which, in its oxidised form, gives brownish or reddish hues. What happens in our British podzols is that the high rainfall and low summer temperatures of upland areas promote a continual passage of water through the soil. As it travels through the soil profile, the water dissolves some of the more soluble minerals from the surface and immediate sub-surface horizons. As the soil water becomes more saturated with iron, aluminium and some less obvious minerals, they are deposited lower down in the profile – as we saw in the Brown podzolic soil. The difference with these podzols is that they have a continuous pale leached horizon (it's called an *E* horizon, for *e*luviated, meaning 'washed out'). Whilst the original Russian ideas of the podzol related to lowland acidic soils on permeable materials like sands, our upland podzols are developed in more loamy deposits like weathered mudstones and shales. Because rainfall is high, and summer temperatures are relatively low, the surface horizon is very rich in organic matter – usually peaty – whilst the subsurface horizon shows evidence of waterlogging – i.e. a grey colour, flecked with orange-brown or yellow mottles. We thus have what I have come to describe, in a highly simplified way, as a "black-white-orange" profile: black, being the peaty topsoil; white, the leached and gleyed sub-surface horizon; and orange, the iron-enriched lower horizon above the bedrock. We call these soils **stagnopodzols**. This is best illustrated by a photograph, or, better, by going out and looking at the exposed soil on the edge of a forest road or moorland track. The lower, orange layer may have an iron pan – looking just like a thin layer of rusty corrugated iron. The iron pan is hard to penetrate, and roots can often be seen running horizontally, or forming a mat above it.

As you might imagine, these soil types are infertile. The nutrients have been washed away into the rivers – eventually to fertilise lower lands. The surface is very rich in organic matter as the low temperatures and wet climate slow the breakdown of plant residues. A specialist type of flora has developed over the years, one that is able to cope with the cool wet conditions. Many of the species found are peat-forming, especially some of the mosses (called *Sphagnum*), and the cotton-sedges. The stagnopodzols on our moorland are called **Hiraethog** series where there is a continuous

Stagnopodzol in mid Wales

iron pan, and **Hafren** series where there is not. Although generally wet, these soils are drier than the surrounding peats and peaty gley soils described below, and the vegetation is usually dominated by mat-grass (*Nardus stricta*), a tussocky wiry grass with rough leaves and blackish flowers. Amongst the large areas of Podzols on our hills, there are numerous rock outcrops, and areas of shallow soils, known as rankers. We have already encountered these in the sections on dunes, and ceunentydd,

Heather flowering on the moors of mid Wales

they occur on several types of parent material, but are best known on rock, and are recognised by the absence of subsoil. They have been described as *A–C* soils, having only a topsoil, or A horizon over the parent material, or C horizon. They are most common over hard igneous rocks which do not weather as much as softer rocks, and on summit convexities, often around small rock outcrops, or on ledges on cliffs. In the uplands they tend to occur in small areas that are not mappable at the 1:250,000 national map scale, and so are under-represented on that map. The mat-grass swards that dominate on this land have only a few associated species, most notably the grass, sheep's-fescue, the tussocky heath-rush and heath bedstraw. A little colour is given by patches of yellow tormentil and the occasional jewel-like flowers of heath milkwort, which can be magenta, blue or white.

If grazing is reduced, and much experimentation is currently being done on the details of this, the acidic *Nardus* grasslands increasingly give way to upland heathlands. I have discussed the desirability of heath in an earlier chapter in relation to lowland heaths and rhosydd. The upland heaths on rankers and stagnopodzols are dominated by heather (*Calluna vulgaris*) and make a glorious sight in late August and September when the purple flowers transform the appearance of the landscape.

Stagnopodzols are prevalent on slopes, whilst flatter areas carry blanket peat. This descriptive name is particularly apt, as the deeper peats form a duvet-like covering over the land once you reach an altitude of 500 m or so. This is because the climate gets cooler and wetter with increasing height above sea level, and, as we have seen, this promotes peat formation. The higher you get, the steeper the slopes that are covered by peat. The peat varies in depth from 40 cm (peats shallower than this are classified as non-peat soils) to perhaps up to 2 m, or even deeper in hollows. In places, where the peat has been eroded away by wind, weather, and the trampling of sheep, an underlying iron pan can be found, suggesting that the peat formed when the climate became wetter and cooler, over a podzolic soil that existed in former times. At least, this has been a popular hypothesis, but there is not enough evidence to say that this is the case: it is possible that the iron pan formed in situ, under the peat. Erosion of peat is very common in some areas, and has exposed the pale subsoils. In places, too, the remains of forest trees are preserved deep in the peat, attesting to a higher treeline in the past, and a much more forested landscape. Like the Stagnopodzols, the peat soils are generally pretty infertile, and the plants that grow there have to cope with this. Two of these upland herbs supplement their nutrient intake by absorbing nutritious compounds from insects: the sundew and the butterwort are known as insectivorous plants. Other plants are just adapted to survive on low rations, and in waterlogged conditions, like the bog-bean, marsh St John's wort and bog asphodel. The Latin name of the asphodel *Narthecium ossifragum* is said to reflect the belief that it caused broken bones in cattle that grazed it.

The tendency for weak bones in these beasts was probably related to the lack of calcium in these acidic soils.

Forest plantations of spruce and pine cover huge areas of the Welsh uplands. They have been fairly successful on the podzolic soils, although their management is problematical. The wet subsoils prevent deep rooting, and trees are subject to being blown over in windy sites. Foresters try to improve the drainage by cutting ditches, and by planting the trees on the ridges created alongside the furrows. They have developed ways of working out the risk of wind throw, and trying to minimise damage by reduced thinning of the crop in susceptible areas, and, in some cases, by early harvesting. Rankers are less suitable for tree planting, again because rooting is restricted by the shallow depth, and they often occur on exposed tops.

Moorland walks

There are many great walks on the moors of Wales. However, it is best to choose a day with at least *some* sunshine. One of the best that I know is a 6 mile circuit from the remote and romantic chapel at Soar-y-mynydd, in the hills of Ceredigion, between Tregaron and Abergwesyn. The drive to the chapel is quite an adventure as the roads are narrow, winding and hilly, and cross the extensive moorlands of the Cambrian Mountains in an area sometimes known as Elenydd. The chapel can be found at SN784533. Following the track southwards from the car park takes you to the secluded farm at Nant-llwyd, and a gateway behind the farmhouse leads to an old track that skirts the northern slopes of the Nant-llwyd valley. This good track winds its way over the moorlands, where you can see most of the features that I have been describing in this chapter. When you reach the watershed, after less than 2 miles, there is a surprising view as the track turns to a footpath, and descends steeply into the delightful, remote valley of Afon Doethie. A couple of ruins, and some ancient crab-apple trees, show that people were living in these valleys in the not-too-distant past, but it's incredibly quiet now.

Turning right up the valley you follow a beautiful path high above the river, with exposures of Manod series and Hafren and Hiraethog stagnopodzols, and wetter peaty gley soils on the lower slopes. The path continues for about 2 miles (3 km) to the vicinity of Ty'n-y-cornel Youth Hostel, but you turn right up the gravelly track. The steep initial climb levels off, and you are in more of the typical moorlands of this area. To your right, about 150 m off the track is the remains of a Bronze Age cairn called Carn Saith-wraig (7 wives). It is now simply a pile of smallish stones on a raised mound. There are many of these monuments on the moors, reminding us that whole communities of people were living, worshipping and dying in these hills long before there were modern roads. Our track soon arrives back at the little Welsh chapel which still holds services from time to time.

CHAPTER 12

Mountains

If you glance at a relief map of Wales, it looks like a wrinkled bedspread, or, better, a piece of aluminium foil that has been crumpled up and then roughly smoothed out. This is not a bad analogy, as the surface of the land has, indeed, been crumpled over millions of years by the collision of the European and African tectonic plates, some 500 million years ago. Since then the mountains created by the collision have been subjected to intense erosion, and were reshaped by glaciers during the last 2 million years. Despite this history of erosion, the mountains of Wales remain impressive, whilst having an attractively approachable quality and pastoral surroundings which draw walkers, cyclists and climbers from all over the world.

A resident of the Himalayas or the Alps might scoff at the idea of mountains in Britain, and will remain unimpressed when consulting an atlas, which shows that none of the Welsh mountains attains the relatively modest height of 1,100 m: there is no chance of suffering from the thin air at this altitude! However, if that same mountain dweller tries a few hikes over our mountains, he or she will soon come to the conclusion that we are justified in calling them mountains. The slopes are steep, and often long, and our westerly situation means that temperatures fall unusually quickly with altitude, and the climate of the mountains can be downright dangerous to the unprepared, even in summer. The rocky slopes and screes and the pastoral and semi-natural vegetation all conspire to make the countryside look mountainous. I remember driving a Nepalese friend from Aberystwyth to Bangor on a superb autumn day, and he remarked (without irony) that it was like the journey from Kathmandu to Pokhara!

There are many differing definitions of mountain, as the character of the land can be independent of mere altitude. The word *mynydd* in Welsh often refers to a relatively low, but craggy eminence, like Y Mynydd on Ynys Enlli, which attains a paltry 167 m, but looks rocky, and quite mountainous in relation to the rest of the island. If you have ever climbed Moel-y-Gest, which overlooks Porthmadog, you may feel justified in calling its rugged 262 m summit a mountain top! Most people know where the Welsh mountains are, but I will run over the surprisingly numerous areas that rise above 600 m, or to older British people, the magical 2,000 feet, where we really do feel that we are in mountain country.

Llyn Cau, Cadair Idris

Looking a little more closely at our crumpled baking sheet, the relief begins to look a bit less random. In the north, in particular, there is a marked tendency for the mountain ranges and the valleys to be aligned from SW to NE: a feature well shown in the Bala fault and its associated valleys from Tywyn via Llyn Tegid (Bala Lake) to Corwen and Wrexham. We begin to be able to separate the ranges and mountain blocks. The best known area is Snowdonia in the north-west, where there are several peaks above 1,000 m. The highest mountains in Wales here present a major barrier to communications, forming a wall along the north coast from the Menai straits to the river Conwy, and separating Anglesey from the lowlands of Ardudwy. The mountainous nature of Wales had had an enormous influence on its history and its language: guerrilla groups were able to disappear into the mountain refuges and withstand the inexorable pressure from colonists from the east for thousands of years. At the same time, the difficulties of crossing the terrain meant that the tribes were never truly united, and local power-bases came and went, depending on the charisma and strategic acumen of their leaders. Each valley developed its own dialect of Welsh, and communication was limited by this.

The high country of the Snowdonia National Park extends southwards, in three dissected blocks, from Snowdon, the Glyders and the Carneddau via the Rhinog hills to Cadair Idris. Each of these blocks has a different character, with the Snowdon area being the most extensive, and almost alpine in its vistas of volcanic range behind range. The Rhinogs are different in that the rocks are predominantly hard sedimentary gritstones, which give very rugged country, much of which is covered with heather. Cadair Idris rises to nearly 3,000 feet (but, at 893 m, doesn't quite make it) and is close to the sea, making it a spectacular and beautiful climb. It resembles the Snowdon massif in that the rocks are mainly igneous.

North-west of Cadair Idris, a high ridge extends across country to the vicinity of Llangollen. Both the Aran and Berwyn ranges here have peaks above 800 m. There are isolated peaks, too, to the north of these, those of the Arenigs which also reach around 800 m. To the south of Cadair Idris there is mainly high moorland, with only a few places rising above 600 m. Pumlumon is

Rugged sedimentary rocks of the northern Rhinogs

Llyn y Fan Fach and Fan Brycheiniog

the most mountainous of these, reaching 752 m, and even having its own corrie lake, the source of the Rheidol, on its northern flanks. The high moorlands continue southwards as the Cambrian Mountains until you reach that glorious east-west ridge formed by the scarp of the old red sandstone. It begins in the east, at the English border, with the Black Mountains, which rise to 811 m at Waun-fach. The Usk separates these hills from the central block, the Brecon Beacons, whose summit is Pen y Fan at 886 m. The Beacons extend westwards to the Carmarthenshire Black Mountain, or Bannau Sir Gâr, although the highest peak is Fan Brycheiniog, right on the border with Breconshire. This peak overlooks the iconic site of one of Wales' most well-known legends, Llyn y Fan.

Mountains, then, are the craggy, rocky bits that emerge from the more rounded moors and represent an irresistible challenge to walkers and climbers. Soils on the mountains are affected by a number of factors: the cool, wet climate favours peat development, as we have seen elsewhere, and soils here are very rich in organic matter. They are infertile because, although weathering is rapid at high altitudes, the processes of erosion and leaching are also accelerated by the extreme climate. There is a lot of bare rock. Certainly the soils are very thin, and may be non-existent (the American soil classification has a category called "not soil"). Bare rocky slopes may be cliff-like on solid rock, or scree slopes that lie below the crags and are constantly changing. In fact, if you go to the rather laborious (and dangerous) trouble of digging through screes, you may find buried soil layers that represent periods of stability when vegetation developed on the scree, and stabilised it so that organic matter began to accumulate. Subsequent movement then buried this layer. The bare rocks themselves have a covering of lichens so that the surface colour is that of the plants and fungi rather than the rock itself. This is why geologists need to break rocks to see their fresh internal colour and structure. Where soils are developed in cracks in the rocks, and on summit hollows or plateaux, they are usually peaty because the high altitude leads to very wet conditions and cool temperatures throughout the year. Hence rankers are the most common soils – thin layers of peat over solid or broken rock. The classification of these *humic rankers*, as they are called, depends on the nature of the rock. Those on sedimentary rocks are called **Skiddaw**, whilst those on volcanic (igneous) rocks are divided into **Bangor** on acid igneous rocks and **Preseli** on basic igneous. Often there is peat which is a little deeper than the 30 cm limit for rankers, so such soils are classified with the peats, despite the fact that they are unusually shallow. The soil illustrated is in this category. The profile is visible because walkers on a popular route up Cadair Idris have eroded the surface, exposing a natural soil profile on the edge of the path. Peaty soils like these often occur on summit plateaux, where I have examined soils at about 800 m. The appearance of these plateaux can be so rocky and stony that it is hard to imagine that there is much soil, but beneath a layer of stones I have found greasy brown-black peat about 35 cm deep. This material provides sustenance for the mountain grasses, sedges and herbs, whose roots spread widely through it. In a few places there is a gleyed mineral layer above the rock, so the soil is classified as a humic stagnogley – this soil has the appropriate name of **Mynydd** series.

Sometimes the summit areas can be relatively flat. The tops of the northern Rhinogs, west of Trawsfynydd in Meirionnydd, for example, are on horizontally bedded coarse gritstones. The surface pavements are bare of soil, as wind and rain prevent its development. The only vegetation and soil is in cracks, that can give the appearance of limestone pavement.

Mountain peat soils on Cadair Idris

Mountain Walks

There are numerous guide books to lead you up all of the Welsh mountains. If you want to see examples of most of the landscapes and soils that I have described in this chapter, I would recommend a walk up my favourite mountain – Cadair Idris – only a 35 minute drive from my home near Aberystwyth. The classic route is a decidedly strenuous 5 mile hike starting from Minffordd, a Snowdonia National Park car park just off the A487 beside Talyllyn, between Machynlleth and Corris (at SH732116). You begin through some fine oak woodlands, climbing steeply beside Nant Cau, which tumbles through the woods in attractive cascades and rapids. The path has been improved in recent years by dedicated teams working for

Cracked rocks on the Rhinogs

the National Park Authority, and is easy to follow. Coming out of the woods, the glaciated valley bends around to the left, eastwards into Cwm Cau, a classic area for the study of glaciated landscapes, visited by many a group of students young and old. Just before you reach Llyn Cau, some 1.3 miles into the walk (it feels like a lot more because you have already climbed some 400 m) there is a large rock shaped like an upturned boat, and about 30m long. If you examine the southern flank of this rock, alongside the path, you can plainly see the scratches made by rocks carried in the ice from the Cwm Cau glacier, some 16,000 years ago. This hard boss of rock probably stuck up so that the glacier flowed either side of it. It is a welcome landmark after the hard climb, and I call it Mars bar rock! Beyond the rock are the dark waters of Llyn Cau, with a forbidding rock wall behind. This is a typical glacially excavated lake, held up by piles of debris left by the molten glacier.

From here you turn to the left (south) and take another steep path that climbs the southern wall of the dramatic glacial cwm, and follows its rim around to the summit of Craig Cwm Amarch. This is often depicted in paintings as the summit of Cadair Idris, as it towers above the lake. There are several precipitous gulleys that lead directly down to the lake, some 340 m below. Unfortunately, it is now necessary to lose height, and continue northwards to Bwlch Craig Cau. If you look at the rocks around you, you will notice that the Bwlch is worn into softer sedimentary shales, marked by frequent white quartz boulders and seams. In fact the glacier has picked out this zone of less resistance, and the shales are evident right through the mid-line of the Cwm.

A steady climb from the Bwlch brings you, at last, to the summit of the mountain. There is a small hut, which you may be glad of if the weather is not good, but if you are lucky there are magnificent views to the north and the west. On a good day you can see Llyn Tegid glistening in the distance, with the town of Bala at its head. You are looking straight up the Bala fault.

For the descent, head for Mynydd Moel, the next peak, a little less than a mile to the ENE. From its summit a rather steep and rocky path descends to a little bridge over Afon Cau, and back onto the woodland path that you started on. If you fancy a little more time on the mountain, and a significantly longer walk, you can keep following the ridge north-eastwards over Gau Graig, and encounter the old Bwlch-y-llyn-bach to Dolgellau road, now a somewhat boggy track, where you turn right back to the summit of the Minffordd to Cross Foxes main road. Shorten the walk down the main road by going through a gate on the right some half a mile below the Bwlch, and follow the pleasant track parallel to the road to return to the car park.

CHAPTER 13

Artificial Soils

Many parts of the National Soil Map are coloured in a rather unprepossessing grey or grey-green. The grey areas are described as "unsurveyed". This does not mean that the soil surveyors couldn't be bothered to work in these areas, it means that they were unable to find enough soil to make the survey worthwhile. They did not, for example, go and dig in people's gardens – so urban areas remain unsurveyed, as do large industrial sites and other built-up areas. This makes sense anyway, as garden soils have been altered so much from the 'natural' state that each garden has its own characteristics (although they usually still reflect the nature of the original soil). Most building sites have been stripped of their topsoils and a great deal of the subsoil. In well managed sites, these are replaced to more or less replicate what was there before, but this ideal is seldom realised. In many examples topsoil has to be imported from elsewhere. The quality of this material is highly variable: I remember being involved in making judgements for the British Standards Authorities on whether material advertised as "good quality topsoil" fitted the description. In one case, material thus described was obviously subsoil, and we were able to identify, more or less, where it came from. Another sample that we inspected was full of pieces of brick, rubble and glass, and had evidently been recycled from some previous housing or industrial development. Larger areas of parks and other open areas in towns and cities were examined by the Soil Surveyors; although I note with some disappointment, that the parks of the Taff valley in Cardiff are grey on the map. Perhaps the City Council was unwilling, at the time of survey, to allow access to a person with a spade!

The grey-green splodges on the map are found mostly around the coalfields, and represent areas where soil has been restored after industrial exploitation of the land – mainly opencast coal mining. A whole secondary industry evolved from the need to replace the soil cover on areas that had been exploited for opencast mining – which is basically a form of quarrying: as techniques and technology developed in the mid twentieth century, coal was dug from ever deeper deposits. In this process, the soils were first stripped and stored in separate topsoil and subsoil heaps. In the early years of the industry this was done in a relatively careless and haphazard fashion, but in the later years, great efforts were made to maintain the topsoil in a healthy condition by avoiding compacting it and taking measures to keep it aerated so as to maintain a healthy population of earthworms and other soil fauna. After several years, when the coal had been removed, the rocky overburden was replaced in the hole and compacted down to prevent subsidence in later years. The subsoils and topsoils were then replaced, drainage systems installed, where appropriate, and a grass crop grown prior to the farming or forestry regime of former times being continued. The results of these activities over the years have produced a mixed bag of soils. I only have experience of the early restorations in Carmarthenshire, where the original land was usually semi-upland on boulder clay. If you have been reading sequentially through this book, you may remember that such land has mainly Surface-water Gley soils: Wilcocks series having peaty topsoils, and Brickfield with non-humose topsoils. These are not the ideal soils to mess about with, and the restorations reflect this. I suspect that much of the topsoil disappeared during the years of working, and the restored soils are generally so poor that they are used only for rough grazing with low stocking rates. The soils are readily identified as man-made by the presence of pieces of coal and other materials, including, in the worst examples, old iron hawsers and other paraphernalia associated with the industry. The extent of this land is depressingly large – it covers over 500 square km in Wales, although a small amount of this total is from workings other than coal. I should emphasise here that modern techniques of restoration are a huge improvement on the past ones (thanks largely to much hard work by soil scientists). In fact, some areas will be more suitable for agricultural use after restoration than they were before.

There is an increasing tendency nowadays to think about restoration designs aimed at uses of the land other than agriculture and forestry. Many old gravel workings have automatically reverted to wetland ecosystems that support a good range of semi-natural plant communities, and these sites can be very good for bird life. Some of the planning applications for gravel and sand extraction aim at restoring the land to lakes and wetlands that are good for wildlife and provide recreation facilities for the local people and visitors.

There is also some *unrestored* land, which has mainly been included in the *unsurveyed* category. This consists of wastes from extractive industries, including coal tips and slate tips. Over time, these man-made substrates have been colonised by plants following the natural ecological process known as *succession*. Mature soils take hundreds of years to form from raw parent materials, but on these relatively newly produced materials we can study the process from its beginning. On slate wastes, for example, algae and lichens fairly quickly colonise the bare rock surfaces, softening the surface of the slaty blocks by producing acidic substances. Lichens are slow-growing combinations of fungi and algae: the fungal element provides the body of the organism, and sends out strands called hyphae to absorb scarce water resources and dissolved minerals. Green or blue-green algal cells occupy the cells of the fungus, and, unlike the fungus, are able to photosynthesise, creating carbohydrates from carbon dioxide and water, thus providing the whole organism with a source of energy.

Lichens and algae on their own soften and roughen the surface, so that small plants like mosses can establish themselves, and a thin moist organic surface covers parts of the rocks – especially those that are partially shaded from the hot Welsh sun. It is possible to peel off this mossy covering, and if you do so, you will see that the mosses are anchored in a black material very much like a thin topsoil. This stuff is enough for the odd hardy grass or small herb to seed into, and once the flowering plants get established, the succession process is well on its way. These hardy pioneers are able to extend roots deeply into the spaces between the slate blocks, where organic and mineral matter have accumulated; some coming from the *in situ* vegetation, and some blown in from elsewhere. Roots obtain nutrients and water from deeper in the pile, and bring them to the surface, hence forwarding the process of soil formation. Even trees are not slow to colonise slate heaps, especially birch and mountain ash, whilst our old friend the rhododendron can, and often does, elbow its way into the act. As well as garnering nutrients from deep in the pile, the plant cover helps to stabilise the tips and prevent erosion and rockfalls.

Lichens and stonecrop on coastal rocks

Mountain ash with heather, grasses and ferns on slate waste at Abergynolwyn, Gwynedd

CHAPTER 14

What did soils ever do for us?

"People have no need whatever to put their houses where they will slide down hill, settle and crack, or be flooded; nor where their homes cannot be beautified with growing plants. It has been demonstrated that what has been learned about soil selection for the many specialties in farming can be used for these other purposes."

These words were said at the beginning of the 20th century by the influential American soil scientist, Charles E Kellogg. Even almost a hundred years ago he thought that we had enough information about soils to substantially improve our lives. Ideas and knowledge have changed enormously since then, and, if we make the effort, we can apply soil wisdom to crucial elements of life on Earth.

I will very briefly discuss just three aspects of this, and the first is **sustainability.** There are soil-free methods of growing things: hydroponics, for example, but they depend on carefully maintained technological systems. Whilst they may have their place in some unusual circumstances, they will never replace the activities of agriculture, forestry, horticulture and gardening on good old natural soil. Hydroponics involves feeding plants with a nutrient solution, but part of the glory of real food is that the taste varies from place to place. This is for a variety of reasons, but one of the main ones is that people have nurtured a relationship with the local conditions to get the best out of the resources available to them. The unique combinations of chemical compounds in the soil will influence taste, texture and character of local produce – just as sources of nectar affect the taste and texture of honey, or the nature of the water in a particular glen determines subtleties of the taste of whisky.

We know so much about our environment now, yet we still build houses on flood plains, or where cracking clays will disrupt foundations. We learn some lessons from disasters like the Aberfan catastrophe of 1966, but we also have a tendency to sleepwalk into repeating them. We need to look at the longer-term consequences of our actions, and to ensure that we do not compromise the conditions that sustain us. Ensuring the ongoing ability of soil to support buildings and to provide food and other necessities of life is essential.

My experience over the years has been that people like engineers, planners and, to a lesser extent, foresters and farmers would like to take soils into account when they are considering projects, but they are unwilling to engage fully with a subject that they don't fully understand personally, and they don't know where to start. My hope is that this book will help give such people the basis for thinking about such things, and the confidence to consult soil scientists who are only too anxious to help.

A second aspect is known as **ecosystem services;** which is to say the ways in which natural systems create conditions that benefit mankind. There is a long list of these, and some are better understood than others, but they include the provision of clean water, the control of flooding, and interactions with climate. The idea is not a modern one – an anonymous 14th century poem lists the ways in which threatened woodlands near Aberdare benefitted the local populace. As well as donating fodder for pigs, and timber and firewood for the householders, they also furnished lovers with trysting places, gladdened the ear with birdsong and ensured that the springwater was clean.

The ecological fabric of the land is much more than a mosaic of managed land with a scattering of semi-natural habitats. It has, by default, provided for human demands ever since we first colonised the land. I say *by default* because this has happened, on the whole, without conscious control by humans. Feedback mechanisms have often ensured that, for example, areas were not over-exploited by early hunters, but not always. It is only now, in the 21st century, that we have realised the extent to which overfishing is depleting stocks in the oceans. Action 50 years ago could have made a huge difference to this. The natural environment of the sea still provides an (albeit depleted) larder for human use, but the land is more like a garden than a larder, and all is not entirely rosy in the garden. North Americans are more aware of this than we are in Europe, having suffered the "dust bowl" of the 1930s. The Great Plains of North America had thick topsoils bound by deep-rooted native grasses. The soil was in equilibrium with the vegetation, and able to withstand relatively long periods of drought. Continual ploughing of these grasslands (rather ominously termed *prairie busting)* had replaced the natural grasses to the extent that drought and high winds began increasingly to blow the topsoil away, and this happened on a vast scale, causing untold human misery. There are many other examples of misuse of the land, including, in Wales, compaction by excessive trafficking or overstocking, erosion caused by ploughing on steep slopes, and even pollution of streams resulting from muck-spreading under wet, or frozen conditions. Knowledge of soil properties and judicious management can reduce or eliminate these activities that so often cause long-term damage to the soil and land.

The third aspect is **climate change.** Scientists from all disciplines are in general agreement that the world is currently getting warmer, and that this is, at least partly, due to the effects of "greenhouse gases" produced by human activities like the burning of fossil fuels. Now soils, throughout the world, are an enormous reserve of carbon, the main element of carbon dioxide, the most important of the greenhouse gases. It is vital that we understand the relationships of soils with the production of $C0_2$ and the importance of maintaining the store of carbon in soils. The digging up, or draining of large volumes of peat for example results, sooner or later, in the release of large amounts of $C0_2$. Exposure of soil when felling forests also releases greenhouse gases into the atmosphere.

This is not the place to give the details of soil-related methods of ameliorating the environment, or to suggest solutions to some of the huge problems that face us, but I would like to make a plea for policy makers, and anyone who takes an interest in these matters (and that should be all of us) to take into account the importance of natural and semi-natural soils in all their variety and complexity. There is an old saying, usually applied in a jocular fashion, that "the answer lies in the soil". Perhaps it is not a joke.

Glossary

Brown Earth
A type of naturally well-drained soil with brown (sometimes reddish) subsoil to a depth greater than 30 cm.

Boulder clay
A usually clay-rich deposit resulting from the passage of glaciers over the land. Boulder clay (geologists and glaciologists prefer the word *till,* of Scandinavian origin) usually contains a mixture of clay derived from the grinding action of the ice, and coarser fragments (sometimes including boulders) carried along by the ice.

Gley
A type of soil with greyish subsoil indicating waterlogging for a significant part of the year. Gley soils are usually wet underfoot except during dry summer periods.

Horizon
Soils have distinctive layers, called horizons. The characteristics and distribution of the layers can be used to classify different kinds of soils. Horizons are conventionally labelled:

A horizon – the surface layer, often called topsoil, usually richer in organic matter, and therefore darker than the layers below.

B horizon – the layer below the A horizon, often called *subsoil,* in which the original geological material has been altered by soil-forming processes.

C horizon – lowest layer of the soil, usually consisting of geological material unaltered by soil-forming processes.

Podzol
A type of soil with a pale grey subsoil which has lost its colour as a result of *leaching,* a process whereby the flow of water through the soil profile washes out the minerals that give it colour. These minerals are deposited lower down in the profile, giving a reddish, iron rich layer below the E horizon. See page 107.

Ranker
A shallow soil (less than 30 cm deep) over non-calcareous rock or some other substrate unaltered by soil-forming processes.

Rendzina
A shallow calcareous soil, usually over limestone, but can be over other calcareous deposits such as sand.

Soil Series
This is the equivalent of the *species* in biological classifications. The dominance in the landscape of a particular Soil Series is what is usually depicted on detailed soil maps. The names of Series usually refer to the places that they were first mapped. Hence *Rheidol Series,* or *Denbigh Series.*

Stagnopodzol
An upland soil with a peaty topsoil over a pale, leached subsurface horizon, below which is an orange iron-rich horizon, sometimes with an ironpan. See page 109 for more details.

Texture
An important property of soils which refers to the proportions of sand, silt and clay in the soil once the stones have been removed. Scientific texts now use the term *particle size class* for this – see page 19 for details. Texture interacts with organic matter content to influence other properties such as permeability and water-holding capacity.

Till
A usually clay-rich deposit resulting from the passage of glaciers over the land. Often called Boulder clay, it usually contains a mixture of clay derived from the grinding action of the ice, and coarser fragments (sometimes including boulders) carried along by the ice.

Gold Under Bracken is just one of a whole range of publications from Y Lolfa. For a full list of books currently in print, send now for your free copy of our new full-colour catalogue. Or simply surf into our website

www.ylolfa.com

for secure on-line ordering.

y Lolfa

TALYBONT CEREDIGION CYMRU SY24 5HE

e-mail ylolfa@ylolfa.com

website www.ylolfa.com

phone (01970) 832 304

fax 832 782